The Palgrave Macmillan Transnational History Series

Series Editors: **Akira Iriye**, Professor of History at Harvard University, and **Rana Mitter**, University Lecturer in Modern History and Chinese Politics at the University of Oxford

This distinguished series seeks to: develop scholarship on the transnational connections of societies and peoples in the nineteenth and twentieth centuries; provide a forum in which work on transnational history from different periods, subjects, and regions of the world can be brought together in fruitful connection; and explore the theoretical and methodological links between transnational and other related approaches such as comparative history and world history.

Editorial Board: **Thomas Bender**, University Professor of the Humanities, Professor of History, and Director of the International Center for Advanced Studies, New York University; **Jane Carruthers**, Professor of History, University of South Africa; **Mariano Plotkin**, Professor, Universidad Nacional de Tres de Febrero, Buenos Aires, and member of the National Council of Scientific and Technological Research, Argentina; **Pierre-Yves Saunier**, Researcher at the Centre National de la Recherche Scientifique, France and Visiting Professor at the University of Montreal; **Ian Tyrrell**, Professor of History, University of New South Wales

Titles include:

Desley Deacon, Penny Russell and Angela Woollacott *(editors)*
TRANSNATIONAL LIVES
Biographies of Global Modernity, 1700–present

Gregor Benton and Edmund Terence Gomez
THE CHINESE IN BRITAIN, 1800–PRESENT
Economy, Transnationalism and Identity

Sugata Bose and Kris Manjapra *(editors)*
COSMOPOLITAN THOUGHT ZONES
South Asia and the Global Circulation of Ideas

Martin Conway and Kiran Klaus Patel *(editors)*
EUROPEANIZATION IN THE TWENTIETH CENTURY
Historical Approaches

Joy Damousi, Mariano Ben Plotkin *(editors)*
THE TRANSNATIONAL UNCONSCIOUS
Essays in the History of Psychoanalysis and Transnationalism

Jonathan Gantt
IRISH TERRORISM IN THE ATLANTIC COMMUNITY, 1865–1922

Deep Kanta Lahiri Choudhury
TELEGRAPHIC IMPERIALISM
Crisis and Panic in the Indian Empire, c. 1830

Glenda Sluga
THE NATION, PSYCHOLOGY, AND INTERNATIONAL POLITICS, 1870–1919

Forthcoming:

Sebastian Conrad and Dominic Sachsenmaier *(editors)*
COMPETING VISIONS OF WORLD ORDER
Global Moments and Movements, 1880s–1930s

Matthias Middell, Michael Geyer, and Michel Espagne
EUROPEAN HISTORY IN AN INTERCONNECTED WORLD

The Palgrave Macmillan Transnational History Series
Series Standing Order ISBN 978–0–230–50746–3 Hardback
978–0–230–50747–0 Paperback
(outside North America only)

You can receive future titles in this series as they are published by placing a standing order. Please contact your bookseller or, in case of difficulty, write to us at the address below with your name and address, the title of the series and the ISBN quoted above.

Customer Services Department, Macmillan Distribution Ltd, Houndmills, Basingstoke, Hampshire RG21 6XS, England

The cable and wireless imagination
Source: Cable and Wireless map, Cable and Wireless Plc., 1942

Telegraphic Imperialism
Crisis and Panic in the Indian Empire, c. 1830

Deep Kanta Lahiri Choudhury
Reader in History, Visva Bharati University, Shantiniketan, India

© Deep Kanta Lahiri Choudhury 2010
Foreword © Rana Mitter and Akira Iriye, 2010

All rights reserved. No reproduction, copy or transmission of this publication may be made without written permission.

No portion of this publication may be reproduced, copied or transmitted save with written permission or in accordance with the provisions of the Copyright, Designs and Patents Act 1988, or under the terms of any licence permitting limited copying issued by the Copyright Licensing Agency, Saffron House, 6–10 Kirby Street, London EC1N 8TS.

Any person who does any unauthorized act in relation to this publication may be liable to criminal prosecution and civil claims for damages.

The author has asserted his right to be identified as the author of this work in accordance with the Copyright, Designs and Patents Act 1988.

First published 2010 by
PALGRAVE MACMILLAN

Palgrave Macmillan in the UK is an imprint of Macmillan Publishers Limited, registered in England, company number 785998, of Houndmills, Basingstoke, Hampshire RG21 6XS.

Palgrave Macmillan in the US is a division of St Martin's Press LLC, 175 Fifth Avenue, New York, NY 10010.

Palgrave Macmillan is the global academic imprint of the above companies and has companies and representatives throughout the world.

Palgrave® and Macmillan® are registered trademarks in the United States, the United Kingdom, Europe and other countries.

ISBN 978–0–230–20506–2   hardback

This book is printed on paper suitable for recycling and made from fully managed and sustained forest sources. Logging, pulping and manufacturing processes are expected to conform to the environmental regulations of the country of origin.

A catalogue record for this book is available from the British Library.

Library of Congress Cataloging-in-Publication Data

Lahiri Choudhury, Deep Kanta, 1971–
   Telegraphic imperialism: crisis and panic in the Indian Empire, c.1830/Deep Kanta Lahiri Choudhury.
       p. cm.—(palgrave Macmillan transnational history series)
   ISBN 978–0–230–20506–2 (hardback)
  1. Telegraph—India. 2. Telegraph—India—History. 3. India—History—British occupation, 1765–1947. 4. Telegraph—Social aspects—India. 5. Social change—India. I. Title.
   HE8374.L34 2010
   384.1095'09034—dc22   2010023837

10 9 8 7 6 5 4 3 2 1
19 18 17 16 15 14 13 12 11 10

Printed and bound in Great Britain by
CPI Antony Rowe, Chippenham and Eastbourne

To
Dhriti and Sheila
My Parents

Contents

List of Figures, Maps, and Tables	viii
Foreword	ix
Acknowledgements	xi
List of Abbreviations	xii
Introduction: 'What hath God wrought!'	1
1 From Laboratory to Museum: The Changing Culture of Science and Experiment in India, c. 1830–56	11
2 The Telegraph and the Uprisings of 1857	31
3 The Discipline of Technology	50
4 Making the Twain Meet: The New Imperialism of Telegraphy	79
5 The Magical Mystery Tour: Cable Telegraphy	105
6 Forging a New India in a Telegraph World: Expansion and Consolidation within India	129
7 The Telegraph General Strike of 1908	157
8 *Swadeshi* and Information Panic: Functions and Malfunctions of the Information Order, c. 1900–12	179
Conclusion	211
Notes	219
Glossary	257
Bibliography	258
Index	271

List of Figures, Maps, and Tables

Figures

1.1	Gentlemanly science	10
2.1	Proposed structure of telegraph lines	36
2.2	The telegraph system that was completed in 1856	36
C.1	Horizontal imagination: The telegraph station at Jask	210
C.2	A different architecture: Marconi's towers	217

Maps

2.1	Uprisings of '1857' and the telegraph lines	30
4.1	The overland routes to India	78
5.1	The Imperial all red line with the British Empire shown in black	104
7.1	A map of the telegraph strike	156

Tables

1.1	Distribution and percentage of private messages sent by Indians from 1 February to 31 July 1855	23
1.2	Breakdown of private messages from May 1856 to April 1857, including numbers sent by Indians until 1857	24
1.3	Regional breakdown of telegrams between November 1856 and January 1857, including those sent by Indians	24
3.1	Breakdown of private telegrams by administrative circle	54
3.2	List of new joint stock companies in Bombay, 1863	58
3.3	Table showing the run on the Bombay stock market in 1863–4	58
8.1	Table showing effects of war on Indian telegraph messages (percentage of decrease)	206

Foreword

If one has to identify one phenomenon above all that subverts the hegemony of the nation state, it is surely technology. In our own era, it is the Internet which is usually summoned up as the clearest example of this tendency, but historical perspective reminds us that there is a story here that long predates the World Wide Web. That is just one of the reasons that the latest monograph in the Palgrave Macmillan Series in Transnational History is so welcome. Deep Kanta Lahiri Choudhury's book examines the importance of the telegraph in reshaping Britain's Indian empire in the late nineteenth and early twentieth centuries. We would do well to remember quite how revolutionary the telegraph was when it emerged as a major tool of communication in the mid-nineteenth century. Regions, countries, and empires could all be brought together virtually with a series of dots and dashes that translated into a means of informing, and of course, controlling societies. In examining this major sociocultural phenomenon, the book moves daringly and excitingly between different levels of examination, forcing us to rethink events that we thought we had understood. One of the most powerful elements of this study is the way that it reworks the story of empire, arguing that the technology of telegraphy was not simply transmitted from the imperial metropole but rather that there was important interaction between the Indian empire and Britain, travelling in both directions. The wide definition of empire here also shows the power of the transnational approach, allowing the analysis of the connections between different parts of the world which had no physical contiguity. One notable element is the reinterpretation of the First World War as a 'communications crisis' in which the non-European world, and the technology that linked it, played a key role. And very importantly, the story told here is not purely one about European elites. We encounter figures such as Kalidas Maitra, author of the first Bengali book on telegraphy in 1855, who was heavily involved in the social reform movements that convulsed Indian society in the late nineteenth century. We also hear stories about subaltern movements from the time of the Uprising of 1857 all the way up to the labour unrest of the early twentieth century. By going down well below the level of nationalist elites, Lahiri Choudhury shows the way that the new technology could reshape politics and society at all levels and link the global and the local. Finally, one should add that it is a pleasure to publish a book so richly based on archival sources gathered in multiple sites both in India and in the UK. This book will serve for generations as the key argument on the importance of the telegraph in shaping one of the most important transnational historical actors, the network that was the British empire.

Rana Mitter
Akira Iriye
Oxford, August 2010

Acknowledgements

A book owes many friends and accumulates years of debt. I thank Sabyasachi Bhattacharya for encouraging my interest in history of communications. I am grateful to Sir C. A. Bayly for his support and guidance. T. N. Harper, S. Scaffer, Francesca Orsini, Robert Fox, Sylvana Tomaselli, and Sugata Bose have been especially kind in their encouragement. Mark Harrison has been supportive throughout the years. A number of grants have supported this work and I thank the Cambridge Commonwealth Trust, the History Faculty, the University of Cambridge, the managers of the Ellen McArthur, Smuts Memorial, Holland Rose funds, and Charles Wallace Trust, and Hughes Hall. A Junior Research Fellowship in Hughes Hall during my doctorate and the Ellen McArthur Prize for the best thesis in social and economic history submitted to Cambridge University in 2002 were great encouragements. I am grateful to the Wellcome Unit for the History of Medicine, and the Modern History Faculty, University of Oxford, for a post-doctoral fellowship. My colleagues and students have been supportive at Jamia Millia Islamia and at Visva-Bharati, Shantiniketan. A number of archivists and librarians in more than one country have facilitated this research and I thank them all. Ian Bolton (AVMG), Phil (Hughes), Kevin (CSAS), Ruth Ireland and Michael Strang (Palgrave Macmillan) have been very generous with their time and skills. The Mitters have been very generous with their friendship and hospitality, as have Sunanda Sen, Sita Narasimhan, and Rita and Oliver Gaggs. Without my parents, Dhriti and Sheila, and my wife, Debjani, this book would never have been finished. Friends and mentors have provided support and comfort over the years and there is no way to thank them except to accept the burden, if it is such, of gratitude.

List of Abbreviations

ASL	Asiatic Society Library, Calcutta
CSAS	Centre for South Asian Studies, Cambridge
GPO	General Post Office Collection, Calcutta
GUL	Glasgow University Library, Glasgow
ICE	Institute of Civil Engineers, London
MSA	Maharashtra State Archives, Bombay
NAI	National Archives of India, Delhi
NL	National Library, Calcutta
NLS	National Library of Scotland, Edinburgh
OIOC	Oriental and India Office Collections, British Library
SB	Special Branch Archives, Calcutta
TCL	Wren Library, Trinity College, Cambridge
WPL	Whipple Library, Cambridge

Introduction: 'What hath God wrought!'

23 ... 'What hath God wrought!'
24 Behold, the people shall rise up as a great lion, and lift himself as a young lion: he shall not lie down until he eat of the prey, and drink the blood of the slain.[1]

The introduction of a general postal service echoed the growth of bourgeois institutions and ushered in the world we know today. The telegraph system fundamentally changed the way the world was viewed in the nineteenth century. The postal service continues, and, the telegraph system too has transmuted into different forms. The giant communications company Cable and Wireless is living testimony to this in our current Internet and mobile world of communication networks.

This book researches the telegraph system of the British Indian Empire, c. 1850 to 1920, exploring one of the single most significant transnational phenomena of the imperial world, and the link between communication, Empire, and social change. During this period the British Indian Empire emerged as a crucial strategic and commercial factor in the telegraph network of the world. The British Indian Empire included present-day Myanmar, Sri Lanka, Bangladesh, Pakistan, and parts of Iran, Iraq, and Afghanistan. It also included the broader sphere of the British Indian 'informal empire' in the Persian Gulf, southern Arabia, the Indian Ocean, and the Bay of Bengal. This book uses the terms India and the Indian empire in this wider transregional, transnational, and transcultural sense. Larger India became a communication hub, and relatively remote points such as Fao and Gwadar were transformed from sleepy townlets and fishing villages in Middle Asia into crucial nodes of the Indian Empire. This electric network serviced trade and communication across nations, territories, and empires from the Americas to the Russias, China, and Australasia including Taiwan, and New Zealand.

This book about the telegraph network and the Indian Empire, c. 1850–1920, explores what was the first transnational electronic communication system, paralleling the modern Internet. The framework of this analysis includes the informal and formal spheres of the broader British Empire, which included parts of Africa, South East Asia, and the Americas, and focuses on issues of information crisis and panic as they recur in governments and peoples. Networks of knowledge, technologies, circulation of commodities and labour, and information monopolies give a new dimension to studies of imperialism, science and technology, nationalist movements, and information crises. This work captures these diverse yet connected spheres as they emerged and transformed the world.

The world of electronic communication, now described as the 'Information Age',[2] began with the telegraph in the 1840s and 1850s. Telegraph technology was the dominant technology until the 1920s, penetrating and permeating other technologies, and prompting changes in timekeeping, language, and the nature and sources of information. The study of the telegraph, paralleling the Internet in popular imagination,[3] including mobile communications technology, is crucial to an understanding of how systems of electronic communication function, and malfunction.

Samuel Finley Breese Morse sent his first telegram from Washington DC to Baltimore in the US on 24 May 1844. Punched on a paper tape, Morse deciphered it at the Baltimore telegraph station to read, 'what hath God wrought? [sic]'[4] This momentous scientific and technological event changed the face of the world. Morse's quotation and his allusion towards divine predetermination and Christian providence behind scientific and technological progress was a recurring theme of the nineteenth-century modernist discourse. This book discusses this theme using examples from early Indian arguments against this propaganda. Visually and physically the telegraph illustrated the power and modernity of the Empire; an extremely important element in the myth of technological empire.

However, in 1855, the first Indian author of a book on electricity and telegraphy argued on behalf of a different, Hindu pantheon.[5] In a book published on steam and the railways, Kalidas Maitra argued that ancient Indians used, manufactured, and possessed the knowledge of steam power.[6] Maitra was constructing an argument, a strategy repeated by many Indian authors, of establishing lost knowledge and forgotten glory for Vedic and Puranic ancient India. This book investigates the myths surrounding the telegraph and shows how they were crucial to the British Empire's propaganda projecting itself as an empire based on Reason, Science and Technology.

The analysis of the first electronic communication network is important for several reasons. First, this study provides an analytical method to comprehend electronic communication technology and society together: themes kept separate in conventional histories of society, and politics, and, of technology. Second, it facilitates the separation between different communication technologies, contributing towards an understanding of the specific social impact of different technologies. Finally, the broader focus of this investigation is communication technology and its intersection with the 'information order'[7] – the variety of debates and sources that manufacture what is valid information in the public arena – to create a particular economy and imagination of politics and information.[8] This book argues that the telegraph network added a significant dimension to imperialism and imperial expansion after 1850, an impetus previously unknown.

The history of communications under British rule has conventionally been written as the history of these systems as public institutions and utilities.[9] Officials writing from within the service revealed postal and telegraphic expansion under their administrations as an expanding process servicing increasing numbers of people, symbolic of the benevolence of British rule.[10] The primary feature of these accounts is their perception of a continuing system of communications that progressively and seamlessly – from the post office to the telegraphs and beyond – provided cheap access to more and more people: an exemplary illustration of the ethic of bureaucratic rule and of civil service. Often written by retired officials in the employ of the administration or commissioned to celebrate significant anniversaries of these institutions, these officials saw them in terms of an undifferentiated naturalistic expansion and an equally smooth transition from imperial control to Indian nationhood.[11]

Two broad traditions of historiography emerge: the first representing an institutionalised and unproblematic narrative of the transition from the colonial to the national, and, the second, investigating various heroic aspects of the success of the ever-expanding metropolitan telegraph technology and its mission of connecting the world for general good. These traditions can be further summarised: first, there is the idea of *colonial legacy*; second, the study of metropolitan and Western-centred research on the telegraph, reflecting the theme of *Western creation* and *peripheral implementation*. Metropolitan laboratories created the universal template and standards, which were then implemented in the peripheries. Indigenous recipients and their reactions, passive receptors of technology transfer except when working as a disruptive force, could then be studied according to their various cultural traits. This book questions these assumptions.

In an article on cable telegraphs, Hunt raised questions about the impact of Empire on British science, and, in particular, British electrical research. Arguing against Pyenson,[12] Hunt concentrated on the role of the Empire 'in changing the shape and direction of metropolitan science itself.'[13] However, his focus on the metropolitan laboratory lead him to write that Britain was the centre of a system where information flowed in and commands flowed out, 'binding the empire more closely together ... [F]rom its position at the centre of its web of wires, Britain was able to ... exercise more direct control over its far-flung empire ...'[14]

First, this work challenges the belief that Britain was the hub of the information network that welded the Empire together: control by Britain, and, by extension, the metropolitan laboratory, is implied in this belief. Second, it investigates the extent of the substantial investment of the British Empire in submarine telegraphy, not only in terms of money but also in terms of ideology, scientific practice, and personnel. Third, it questions the idea that perfect information is possible, that it is a common good, and the picture this ideology painted of endless peace, prosperity, and progress in a global economic and information free market. The telegraph was represented as neutral, accessible to all, and as a means to peace and progress. This was *laissez faire* at its most utopian. Finally, this book shows how daily life and its representation, along with the telegraph, changed in fundamental ways between 1850 and 1920.

Hobsbawm argued for the 'long nineteenth century', stretching from about 1776 to 1914. It witnessed a double revolution: one that began undermining and destroying monarchical rule and the industrial revolution, which saw the rise of an imperial industrial west dominating the rest of the world.[15] Lenin saw Imperialism as the highest or more literally, the 'latest' stage of Capitalism.[16] In writings from 1880s to the 1920s the driving force of history was perceived to be the story of scientific, technological, ideological expansion of the 'West', and more dogmatically, of Capitalism: histories of Europe, European money, and the white Anglo-Saxon race.

Fieldhouse pointed out that Lenin's model explaining imperial conquests in the second half of the nineteenth century did not justify the profit motive of capitalism. Conquests did not yield as much money as imperial conquests had in earlier times.[17] Empires were also increasingly becoming almost too costly to control: they yielded less and demanded more. Stokes pointed out that Lenin was diagnosing the causes of the First World War and not of late nineteenth century imperialism or the 'scramble for Africa' between European imperial powers.[18]

Gallagher and Robinson suggested that the debate on imperialism was far too Eurocentric.[19] The British Empire expanded not just because it was more profitable to do so but because of significant cultural, political, informational, and social imperatives at the frontiers of formal empires. It was this conjuncture between the local and the transnational which went beyond any particular motive dominating imperial expansion on a global level. The mentality and immediate reactions of the official European 'man-on-the-spot' needed to be reconfigured into histories of Empire. According to them, the variety of European interactions with the colonised locality, province, and region had to be understood to return to histories incorporating and highlighting historical agency, and indigenous and imperial collaboration, at the outposts of Empire.

Bayly argued that information panics, shortages, and crises at the margins of the British Empire forced the it to expand informal and formal control.[20] The fear of an antagonistic unknown just beyond imperial frontiers or the recurrent nightmare of imperial Russian penetration regularly forced the Empire to annex territories that were not apparently profitable or hospitable. This added a further twist to Gallagher and Robinson's explanations of late-nineteenth century imperial expansion, that is, it was not economic profit but a spectrum of crises and emergencies at the peripheries that forced imperial expansion.

The sense of underlying continuity prevailed among historians who argued against the notion that there was anything new about imperialism after the middle of the nineteenth century. They specifically targeted the term 'New Imperialism', which came into popular usage in Britain after 1880. New imperialism, which was an important concept to its contemporaries, is not even mentioned in the *Oxford history of the British Empire*.[21] Koebner studied the evolution of the word imperialism but dismissed the term New Imperialism as jingoistic: it was a political slogan and there was nothing new about imperialism.[22]

This book argues there was something new about imperialism after 1850 and that British popular perception was shaped through the experience of the first global electronic network. After 1850, a second phase of imperialism can be observed: a New Imperialism driven by technologies such as telegraphy. Time, space, and language changed, along with nations, people, empires, and information. The book suggests that the term New Imperialism can be usefully deployed in the context of late-nineteenth century science and technology, and, imperial expansion. It revives the notion of New Imperialism in association with the notion of an Empire of Science and returns to its particular significance in the eyes of its contemporaries.

This book extends Bayly's concept of 'information panic'[23] and Anderson's 'imagined community'[24] to formulate the concept of an imagined state for which the telegraph was crucially important. The book studies the continuous interplay between metropolitan needs and peripheral necessities to argue that there was a systemic technological crisis facing the British Empire towards the end of nineteenth and the beginning of the twentieth century. Indeed, the very system of increased demand for information and circulation led to a heightened and more apparent degree of state surveillance that provided impetus to information panics.

This research studies the introduction, experimentation and fruition of telegraphy in the wider Indian Empire to understand its impact on indigenous society. It demonstrates that social and technological practice intermesh in the case of telegraphy. Communication systems and the telegraph in particular, are the means of production of time and space. They do not 'emerge' nor are they exclusive but they are produced and transformed by means which, though apparently obscure and far too every day, regulate our lives to a great extent. The book argues that the British Government, in part fashioned by the telegraph, became more invested in and more vulnerable to different kinds of information flows. In its quest for stability and control it did not realise the extent to which uncertainty, instability, and information panics were inherent in the system. The telegraph had crucial political dimensions and was to form the basis of an intricate and inherently fragile world order based on issues of sovereignty, territory, the apportioning of rates and tariffs, questions around who would build and work the telegraph, and how it could be rendered safe from interruption.

Chapter 1 describes the first experiments in India and the potential and implications of the original project, including Indian responses to the new technology. It examines the changing social, political and scientific economy within which the telegraph was first experimented with and then introduced into India. It delineates some of the problems raised by previous research on technology transfer under imperialism and questions of technological diffusion in indigenous societies. Previous studies of the early history of the telegraph in India have insufficiently addressed the question of the role of the state in transfers of technology between the industrialised West and the underdeveloped East in a colonial setting. Yet the role of the state and scientific societies in the public sphere in diffusion of scientific and technological knowledge was central in the case of industrialisation in Europe,[25] and the chapter analyses the function of the Asiatic Society in locating the changes occurring in the nature of science and scientific practice in India. Chapter 2 examines the impact of telegraphy on the mutinies and uprisings of 1857–8 and the impact of these events on telegraphy in India and telegraphy at the global level. The reconstruction of the telegraph in India is described in Chapter 3. It shows how technological discipline and its subversion functioned at the micro-level, below the level of the emerging idea of the nation state. It also deals with issues of staff discipline and telegraphic practice, and the impact of telegraphy on aspects of everyday life. Chapter 4 illustrates the imperial and transnational context of telegraphy. It analyses overland telegraphy and India's emergence as a crucial strategic hub in the telegraph network. It examines the intersection of telegraphy with politics on the frontiers of Empire and the informal empire of telegraphy. This chapter demonstrates that there was something substantially new about imperialism after 1860 as telegraphs began to encircle the globe, and European powers such as France entered into a race with Britain. For Britain, they were to be the 'crimson lines of kinship' that would bind the British Empire together. For example, Cecil Rhodes proposed to take the telegraph line from South Africa to join the lines in Egypt, traversing the heart of Africa. The chapter reviews the rise of cable monopolies and their relationships with India and the British Empire, especially in terms of information monopolies. Chapter 5 discusses the imagination of telegraph technology in the late Victorian era and examines various monopolies of information and its physical transmission being forged through the telegraph system. It analyses the standardisation of telegraph technology and the basic problem of maximising the number of messages versus the need to meet the minimum operating costs of the telegraph. Chapter 6 discusses labour problems that emerged and which could, and did, arrest communications. Strikes travelled along the nerves of the telegraph, leaving states, regions, and networks helpless as the communication system betrayed it. The chapter explores in detail the telegraph general strike of 1908, which stretched across different nodes of the British Indian Empire. Chapters 7 and 8 delineate the interconnectedness of telegraph and society. They discuss the increasing levels of state surveillance that the new communication systems facilitated and demanded. These chapters also engage with the issue of the kind of information order the telegraph encouraged and generated, through a discussion of rates charged on press messages, the increasing velocity of circulation of information on a global scale, the snowball effects of panic, and the kind of information promoted by the British government in India because of their anxieties over access to and control over information. Finally, the book introduces the concept of an 'imagined state' and its 'entanglement' with 'information panic', and argues that the political crises and the panics of the first decades of the twentieth century were not just disparate phenomena but reflected a mutually reinforcing technological and social crisis with a transnational set of causes and influences illustrative of transnationally networked societies. In the present age of rapid and competitive technological advancement the study of communication history specifically that of telegraphy becomes essential for our better understanding of the nature of the development involved. This book provides a step towards this understanding.

Figure 1.1 Gentlemanly science
Source: Boileau papers, CSAS, Cambridge University

1
From Laboratory to Museum: The Changing Culture of Science and Experiment in India, c. 1830–56

The nature of the colonial state and science and technology in India underwent fundamental changes over the first half of the nineteenth century.[1] Though the trends towards particular kinds of changes were present before the uprisings of 1857, the process accelerated sharply after those events. In this chapter, a discussion of the role of the Asiatic Society, the first and foremost scientific and literary association in British India, is used to examine the relationships between the diffusion and generation of knowledge, changes in public subsidy, the changing nature of scientific and technological practice, and the emergence of a 'colonised' science and technology. This chapter also analyses William Brooke O'Shaughnessy's early experiments with telegraphy between 1836 and 1839 and the telegraph system he established in India in 1856. While O'Shaughnessy was experimenting with telegraphy at the same time as Samuel Finley Breese Morse and others, the East India Company's government did not order O'Shaughnessy to begin construction until 1851, almost ten years after the formation of European telegraph systems. By 1856, a telegraph network of over 4000 miles linked Peshawar, Agra, Bombay, Madras, and Calcutta to the military cantonments and important European settlements in the interior. This chapter examines the features of the telegraph system O'Shaughnessy built, and situates both the system and its inventor within the context of the time.

The Asiatic Society and the changing nature of scientific practice

The Asiatic Society, established in 1784 in Calcutta, was the first and premier scientific and literary association in India. It was founded by

Sir William Jones to facilitate individual production of universal knowledge, that is, anyone situated geographically anywhere can contribute to a universal pool of knowledge[2]. For example, the transactions of the Royal Philosophical Society stand testimony to this endeavour. As the dominant association for the dissemination and production of scientific and technological knowledge in nineteenth-century India, it offered a focus for debates surrounding the nature of knowledge-production under the rule of the East India Company. In the first half of the nineteenth-century, most members of the Society were government officials, and scholars have argued that this made the Asiatic Society an integral part of colonial governance, and inherently 'orientalist'.[3] Writers, like Partha Chatterjee, have dismissed the society's original charter which aimed to produce knowledge from India that had universal relevance and application: 'The oldest and most prestigious institution of colonial knowledge was the Asiatic Society of Bengal. ... Indeed, the main achievement of the society was to make available for modern European scholarship the materials of an oriental civilization.'[4]

Others have argued that the Asiatic Society was devoted to the production of scholarship and knowledge unaffected by government policy. Echoing this second view, Kejariwal wrote that his book 'hopes to establish ... that the world of scholarship and the world of administration were worlds apart ... till 1836 the Society neither asked nor received any substantial help from the Government.'[5] The analysis presented here differs from these approaches. First, it demonstrates the substantial financial reliance of the Society on government subsidy throughout much of the period under study, materially enmeshing the world of scholarship and administration. However, it also argues that to dismiss the Asiatic Society as essentially orientalist is to ignore the changes in its role in the production and dissemination of knowledge. Kejariwal noted, but only in passing, that between 1832 and 1838, 'the very character of Indological studies changed, with greater emphasis on ... visiting archaeological sites, collecting specimens and coins and deciphering ancient scripts of India.'[6] Similarly, Drew recorded that 'from the time of Colebrook on, Sanskrit studies and their fruits were to be the property not of literary men but of professional scholars and philologists.'[7] The following analysis counters notions of seamless continuity with a study of changes occurring within the Society and in the larger context of scientific knowledge production in India.

From 1790 until 1820, the Society published only a few expensive and heavy quarto volumes called the *Asiatic Researches*. More devoted to literary pursuits and the publication of long and finished works, it was unsuited to the growing scientific and technological ferment in the first half of the nineteenth century, which made rapid publication of discoveries, announcements, and notices, imperative. *Gleanings in Science*, launched in 1829, accommodated the demand for short notices in the field of science and technology.[8] This series received a postal subsidy from the government and many government departments and officials were its subscribers. In 1831, the Asiatic Society stopped publishing *Asiatic Researches* and *Gleanings*, and launched a new series called the *Journal of the Asiatic Society*. A significant function of an association in the public sphere is to communicate, and the postal subsidy that the Asiatic Society received from government, crucially facilitated the dissemination and circulation of its findings. Another significant contribution from the government was through subscription. The government and its officials and departments ordered a regular number of copies in advance. These subsidies were vital to the Society's activities, especially in the sphere of science and invention, which, because of rapid announcement could lay claim to equality and universality alongside Europe and the US. This is perhaps the reason behind the prominence of the Physical Committee within the Asiatic Society during the period 1825–40. This committee facilitated investigations, notices, and experiments in science, technology, and medicine. Events soon ended this attempt to create a roving metropolis.

The changing fortunes of the Asiatic Society, British India's main research association, were entangled in the conflict between Thomas Babington Macaulay's 'Anglicist' and Henry Wilson's 'Orientalist' factions over the direction of government education subsidies. The East India Company sanctioned an annual sum to support educational institutions and the promotion of useful knowledge through translations in Persian, Sanskrit, and vernacular languages. When this came up for renewal, two factions of government officials formed. One camp wanted to continue the old system as the best means to promote effective knowledge among Indians while the other insisted on English education, arguing it to be the only language for transmitting European knowledge. This clash would not only shape the future of the Asiatic Society but would also have an enormous impact on the future history of the Indian subcontinent. Increasingly, the Government of India and the Asiatic Society engaged in bitter debate over 'English' versus 'Oriental' education. The senior members of the Asiatic Society were the officials who supported education in Sanskrit and Persian, and the continuance of grants to the Sanskrit College, the Calcutta Madrasa, and similar institutions in Varanasi and Delhi.

The victory of Macaulay's Anglicist faction had a long-term impact on education in India: mastering English came before physics or chemistry. Given this context, in 1834, the government withdrew the postal and publication subsidy that the Asiatic Society had enjoyed since 1829.[9] The Asiatic Society was already in turmoil because of the financial crash of European and British Agency Houses in 1829.[10] Significantly, between 1829 and 1832 there were no publications.[11] The withdrawal of government subsidy was not an isolated case but a part of the systematic cutback in support of public sphere associations devoted to experimental science and the diffusion of useful knowledge during this period.[12] It would, therefore, be simplistic to argue for entirely separate worlds of scholarship and administration, indeed, politics and experience, and research.

Between 1830 and 1855, the Asiatic Society changed from being the premier association for the dissemination and acquisition of knowledge in India to one for the preservation and a depository of knowledge. It was not essentially 'orientalist' to begin with, but became increasingly so. While these trends were maturing throughout the first half of the nineteenth century, the withdrawal of the postal subsidy in 1834 was the decisive act for the future of the publications of the Asiatic Society. A statistical map of the number of articles on different subjects reveals that ethnography, philology, statistical, and religious studies began to dominate the journal after 1834. Said commented that 'with the rise of ethnography ... as demonstrated in linguistics, racial theory, historical classification. ... There is a codification of difference ...'[13] Older subjects such as botany, pure mathematics, and geology, declined in importance, and articles on science showed a general stasis. Significantly, there was an increase in the serialised publication of statistical data, for example, concerning the numbers of storms in the Bengal region, and, tidal and climatic tables.[14] These were not original, nor necessarily relevant to the kind of research the Asiatic Society originally claimed to promote; statistical compilations replaced original scientific research. From around 1840, the Asiatic Society turned towards institutionalised oriental and antiquarian scholarship, the product of generations of scholars, and to the upkeep of its museum. This represented the classic orientalist phase, and a shift from the original charter of Sir William Jones.[15] Knowledge now was to be increasingly about India, serialised, and orientalist.[16]

The ideology informing this shift viewed India as in need of preservation both for itself and from itself. Simply put, the view over this period changed from regarding the colony as a laboratory to viewing it as a museum.[17] It was also a period of transition from original scientific enquiry to state-controlled technological enterprise.[18] During these years, the Asiatic Society appealed for financial assistance from the government to set up its museum.[19] It received a regular amount for the museum and for the salary of a curator.[20] It also received a grant to publish a definitive series on translations of works like the *Upanishads*.[21] The government gave the Society all its oriental collections for publication. The Asiatic Society launched the long-term project of the *Bibliotheca Asiatica* in 1845. The Society's dependence on government subsidy was even more acute in this period because of decline in its funds, publications, and membership subscriptions. In 1835, the society had 306 members; by 1838 it had dropped to 126, and by 1845 had reached as low as 119. Between 1845 and 1860, the number of members varied between 130 (1851) and 242 (1860).[22]

The number increased after 1860, possibly because being a member of the Asiatic Society had by then become a matter of prestige. The Society's history illustrates how the initiative and locus of scientific experimentation shifted to Britain after 1858, though the trends were visible before that date. The Asiatic Society's changing fortunes reflected the transfer of experimentation, production, and the generation of technology from India to Britain. In similar processes, in 1828 the government forbade any involvement of officials with the media. In 1830, it prohibited the use of Indian clothes during the hours of duty and in public. The gaslights, steamships, and the railways, all went to metropolitan patentees. The initiative for these enterprises originally came from private individuals in India but the government concentrated on tightening its lines of control over enterprise in the colony.

Edney commented on the drive to centralise and institutionalise the Great Trigonometric Survey (GTS) after 1830: 'the GTS lost its character as a personal institution … and was transformed into a *proper* institution' (italics mine).[23] Institutionalisation, in this case, stifled original enterprise. By the 1880s, the Society, renamed the Asiatic Society of Bengal to distinguish it from the Royal Asiatic Society founded in London, fought a rearguard action in the international arena. Strong trends towards the centralisation of the imperial state combined in the 1830s with the increasing regulation of opportunities for initiative permitted to government employees and those in non-official sections. The repeal of press censorship in 1835 is an example of such control.[24] It is important to note that the rise in postal rates for newspapers coincided with the introduction of registration of printing presses.[25] This stemmed from the state's changing strategies of censorship, the exercise of invisible controls, manipulation through subsidy, and moral persuasion.[26]

The colonial state was not strong in this period but these trends became important as the state gained in power. In the case of the telegraph, similar strategies of tariff manipulation promised control over the traffic in information.

The changing nature of scientific culture

William Brooke O'Shaughnessy, born in Ireland in 1809, graduated in medicine from the University of Edinburgh. He sailed to India as an Assistant Surgeon in the employ of the East India Company in 1833.[27] His name was among the list of officers attached to the Bengal Military Establishment in 1834.[28] O'Shaughnessy's experiments and competence at public demonstrations in the salons of Calcutta, especially Governor General Auckland's select gatherings, brought him into favour with the establishment. He received financial support for his experiments from donations; Governor General Auckland, for example, made a substantial contribution towards the construction of a powerful battery. In the same year he was appointed Deputy Assay Master at the Calcutta Mint, one of the key scientific positions in British India, through a 'public notice'.

O'Shaughnessy's career reflected the diversity of training that he received while studying medicine at Edinburgh University. The Scottish Enlightenment and the influence of Common Sense philosophy in Scottish educational institutions saw the syllabi, including the science curriculum, adopt a holistic approach: all aspects of philosophical debate were combined with a detailed knowledge of the body, and of physics and chemistry.[29] Although he arrived in India as a member of the Military and Medical Service, O'Shaughnessy soon became second in-charge of the Calcutta Mint. The mint produced and converted coins; *Pagodas*, a currency common in southern regions of India and other currencies coming into the mint had to be tested for levels of purity before conversion. O'Shaughnessy held an office that required a thorough knowledge of chemistry as well as being an important political post. He was in charge of converting diverse gold currencies into a standardised gold coin issued by the East India Company from its mint and bearing its seal. O'Shaughnessy's success was due, in part, to his ability as a demonstrator. He worked not just within the laboratory but also lectured and performed experiments in public and at select salons. This tradition of entertaining science was changing during his lifetime to a more institutionalised definition of science, scientist, and scientific practice.

Working within the traditions of alchemy and the transmutation of elements, O'Shaughnessy saw electricity as a means of bridging the divide between the animate and the inanimate world, using basic Aristotelian elements such as earth, water, and fire. This research laid the foundation for his studies in conductors and conductivity. O'Shaughnessy's experiments were a search for the unity of the vital, or living, and the physical, or electrical, forces. In this search for unity, following Don Francisco Salva, O'Shaughnessy located the fact and analogy of such correspondence in the human body. He experimented with 'Sympathetic Flesh Telegraphy' that involved electric shocks on human bodies so that they could communicate through varying impulses, and though a few of his experimentees were debilitated, he recorded that 'the delicacy of the impressions of touch transcends the sensitivity of all other senses. The eye and the ear are liable to distraction ... while the attentive touch knows no interruption.'[30] In his conceptualisation he combined a belief in extreme central control, substitution by and involvement of the indigene in the colony, and a paternalistic and strongly reformist government. O'Shaughnessy was a pioneer, experimenting with water as an electrical conductor, the usefulness of marijuana and opium for the amelioration of extreme diseases such as tetanus,[31] and with photography between 1837 and 1838.[32] He was a member of the London Electrical Society, founded in May 1837,[33] and was made Fellow of the Royal Society in 1843 and officially declared a Surgeon in 1848.

O'Shaughnessy was associated with the administration of the Asiatic Society from 1837 to 1850. A leading figure on the Physical Committee, he contributed his first article to the *Journal of the Asiatic Society of Bengal* in 1834.[34] Between 1835 and 1850 he was President of the Physical Committee, Joint Secretary, General Secretary, and Vice President of the Society.[35] A study of O'Shaughnessy's decisions at the highest level of the Asiatic Society reveals much about the choices he made later in his career as Director General of the Indian Telegraph Department. It outlines his experience of scientific research in the colony and the government's support for scientific innovation in the first half of the nineteenth century. O'Shaughnessy was part of the important changes between 1830 and 1850, participating in the process of centralisation of government, in the institutionalisation and professionalisation of scientific enterprise, and in the hardening of British attitudes. Over the 1830s the state defined the role and activities of Europeans and employees in the public arena. Given his presence at the senior levels of the Asiatic Society and his official position at the mint, O'Shaughnessy was at the centre of these events. He was one of the key figures guiding the Asiatic Society through this period of change.

O'Shaughnessy's role during the debate between Anglicists and Orientalists, reveals the shifting context he was experimenting in and the limits of his originality. His induction into the scientific establishment of British India was strategic: Macaulay and his faction needed people to support and man the new institutions it was setting up; O'Shaughnessy was a recruit. In 1835, the Committee appointed him Professor of Chemistry at the newly formed Medical College in Calcutta. This was at his own request to the Government of India, in the form of an application sent through the influential Committee of Public Instruction headed by Macaulay.[36] When the Government shut down the physical and chemical science lectures by A. Ross at the Presidency College, Calcutta, O'Shaughnessy proposed that its students attend his lectures at the newly established Medical College.[37] He would exhibit this ambivalence towards Indians in his dealings with the telegraph establishment. He was very much a part of the new order brought in by Governor General Bentinck, and Macaulay, yet retaining the vestiges of an open-ended cultural society. W. N. Forbes, O'Shaughnessy's senior at the mint and architect of St Paul's Cathedral in Calcutta, suggested that the telegraph project be given to English and European patent-holding firms.[38] This was an example of the conflict between 'savants' and experts; Forbes belonged to the Royal Engineers, while O'Shaughnessy belonged to the Medical service. Ghose ignores O'Shaughnessy's role in the Asiatic Society, arguing that '...liberal and firm government support induced him [O'Shaughnessy] to come out of the laboratory and launch a grand technological enterprise'. If the government rejected Forbes' proposal what it allowed was a rigidly controlled department of state to run the enterprise.[39] Thwarted by the government from pursuing his original scheme of subterranean lines, he became pioneer of submarine telegraphy.

The first line and the final system

Between 1834 and 1837 there was a race for the patenting of the best telegraphic device. Morse, O'Shaughnessy, Wheatstone and others were inventing separate versions of telegraphy and were harnessing electricity for different purposes. O'Shaughnessy joined this race but had no right to patent any device not constructed in Britain. His use of original methods and local resources are detailed in the following sections. The total distance from Calcutta to Kedgeree is 82 miles, and O'Shaughnessy used non-insulated iron rods weighing a ton to the mile, and 3/8 inches in diameter, to construct telegraph lines. This followed from his first experiments and was noticeably different from the systems used in America,

England and Europe. Iron rods were relatively immune from 'gusts of wind or ordinary mechanical violence.' He argued that they could not be tampered with easily and their mass allowed a free passage to electrical current that atmospheric turbulence did little to disturb. Most importantly, this innovation made redundant the entire baggage of wires, expensive winding instruments required for keeping the wires taut, insulators and non-conductors. Even with transport costs, it was cheaper than the wire system; and there were, of course, elephants. Iron, as previously mentioned, was easily available in India and local forges were numerous. He used bamboo for the telegraph poles. Its flexibility and low cost made a difference of Rs. 472 per mile when compared to the cost of building and maintaining timber posts using, for example, *Sal* wood. He recorded that the 'over ground system on the plan I have followed presents the great advantages of rapidity of construction, exceeding cheapness, and immunity from storms, lightning and wanton injury ...'[40]

O'Shaughnessy redesigned the telegraph instrument so that it could work 'in all weathers without danger of interruption.' He designed a battery made of 12 to 20 pieces of platinum wire with zinc plates and claimed that 'it suffices to work our lines and instruments through heaviest rain and most violent storms.'[41] He was assisted in the construction of the instruments by the workshop run by Messrs Grindle and Crible in Calcutta.[42] O'Shaughnessy stated that he could provide 'all stations with complete sets of instruments of every kind, battery, reverser, telegraph and alarm, with dozen reserve telegraph instruments, for less than a hundred rupees.' After completion, the average cost per mile of the line came to about Rs. 452. He concluded that he could construct the line in the future for Rs. 350 per mile for a single and Rs. 650 a mile for a double line.

River crossings and the building of offices would be separate charges. The initial high cost was due to the experiments that O'Shaughnessy conducted on almost every aspect of the telegraph. In 1853 O'Shaughnessy's signalling instrument was 'reduced to such a condition of simplicity that when deranged they could be set right by mere schoolboys' without which he thought 'regular and sustained correspondence would be totally impracticable.'[43] Accordingly, because of the rapid polarisation of the magnetic needles, especially during the monsoon, he successively tried and rejected the English Vertical Needle[44] instrument and the American Dotter.[45] The thunderstorms common in Bengal repeatedly put the needles *hors de combat*. What happened was that the magnet in the needles became permanently polarised because of the electricity in the air and stopped activating the markers. O'Shaughnessy's design replaced these instruments with the Single Needle Horizontal telegraph instrument in 1853. He recorded that it was 'now in use in all our stations and with which we work in all weathers without danger of interruption.' He also simplified the electric current reverser in '... solidity and strength, until it now totally differs from any instrument of the kind used elsewhere'. His telegraph instrument was simple enough to be built by the signallers at the cost of Rs. 3 including a margin of profit after construction. O'Shaughnessy records how he was 'driven step by step to discard every screw, lever, pivot and foot of wire, and framework and dial, without which it was practicable to work.'[46] This was not simply a case of downward adaptation of technology from Europe to Asia[47] but also had the potential for substitution and spin-off effects in India.

Telegraph personnel before the closure of experiment

There were two aspects to O'Shaughnessy's early experiments: first, his ability to substitute technology and, second, his courage in striking out on his own. As a result, links were forged with local and indigenous manual labour, artisan or skilled craftsmen, and local manufactories and workshops in terms of spin-off effects. Instruments, including his camera, were designed and manufactured in Calcutta, employing diverse skills such as watch making, glass making, carpentry, and, smiths and foundries. O'Shaughnessy established several workshops after 1856 to build, maintain, repair and replace instruments and machinery: the main workshop was in Bangalore.[48]

O'Shaughnessy originally proposed to employ five Bengali boys with good knowledge of English. He trained them personally for a month at the training school he set up in Alipur, not far from the Governor General's residence in Belvedere. He sent some of them to Diamond Harbour and at least one of them died of malaria: the rest of the batch refused to leave Calcutta. O'Shaughnessy then turned to young Eurasian and European boys at the La Martiniere School in Calcutta, the oldest being 18. The list of establishments for 1855–6 had a preponderance of Eurasians with a few new Indian recruits such as Anunto Ragoba, Ramchunder Dajee, Luximow Baljee Rao, Vittoo Luxmonjee and Govind Viswanath.[49] The experimental part of the system used the resources of two leading mechanics in Calcutta.[50] O'Shaughnessy's original design incorporated important elements of indigenous metallurgical traditions and Indo-European skills established in Calcutta. These skills, the use of trained labour, and the workforce involved in a village forge are often ignored in accounts of indigenous participation, as is the daily labour of whole families on the actual construction. From curious crowds and farseeing businessmen to metal workers and *mistris*, a broad spectrum of the indigenous population was very much a part of the success of O'Shaughnessy's telegraph project.[51]

Of the original recruits, those who survived in the department were, first, Sheeb Chunder Nundee, and second, Shreenath Newgee (Shrinath Neogy?). The rest of the indigenous establishment consisted of peons, jemadars, lascars, clerks and sweepers; their names rarely appear in the official record. Sheeb Chunder Nundee, who remained with O'Shaughnessy from his days at the Mint, was one of the most remarkable recruits. He had helped in the original telegraph experiment and was an important figure in the establishment[52] until at least the second half 1860. In some ways, Nundee is one of the earliest and best examples of indigenous knowledge in telegraphy. Recruited from the mint when he was 25 years old, Nundee was in turn Head Signaller, Accountant, and Manager of the Alipur Central Telegraph Station in the first telegraph establishment of 1851. His salary was Rs. 75 per month.

By 1852, Nundee was an Inspector, First Class, with a monthly salary of a Rs.100 and a travelling allowance of Rs. 40. His rank was just below O'Shaughnessy during the early phase of the telegraphs. He was in charge of the department with Lieutenant P. Stewart when the uprising of 1857 broke out: O'Shaughnessy was in England buying machinery for the telegraph in India. Nundee is also a part of the picture of the 'able assistant' that contained the indigenous informant and contributor in the post-Macaulayan age, that is, the Indian could advance as far as being the 'loyal able assistant', before being subsumed within the bureaucratic and European hierarchy. Subsequently, more indigenous personnel were inducted but they were no more than skilled operators, quite ignorant of the science involved. There was also growing Eurasian and European dominance in the telegraph staff. Scientific and technological experimentation, invention, and the entire production process was transferred to Britain, stopping all spin-off effects in India. As a result, the colonies became sites of adaptation and modification of what became a predominantly metropolitan technology.

Indian use of and response to the telegraph system up to 1857

The response to the 82-mile line that was opened to the public on 4 October 1851 was phenomenal. Most of the government attended: the members and secretaries of the government, members of the Military Board, Superintendent of Marine, members of the Chamber of Commerce and the Trade Association, Managing Director of the Railway Company, and the Consulting Engineer of the Government. There were also 'the community of Calcutta who in hundreds have visited the office.'[53] The Telegraph Act XXXIV of 1854 prevented the general public from visiting telegraph offices. Large-scale construction began on 1 November 1853 at 20 different places and in three rapid phases. A temporary or 'flying' line of iron rod and bamboo was set up for immediate communication and mainly for military use. This was quickly replaced with permanent posts, wires and insulation. These were all single lines and the phase of building double lines did not begin until after 1857. One of the exceptions was the double line to Barrackpur completed in 1855.

On 24 March 1854, as Superintendent of the Electric Telegraphs in India, O'Shaughnessy wired the Governor General conveying his respects from Agra, and Dalhousie telegraphed his congratulations in reply. In less than five months since construction began, 800 miles of telegraph were now functional. The rapidly expanding East India Company's possessions demanded equally rapid expansion of the telegraph system. By 1856, over 4000 miles of telegraph were functioning. Dalhousie had gone on to annex Pegu, the Carnatic and the rest of Central India. By the end of 1855, the entire line from Sagar Island, on the coast near Calcutta, to the key military cantonments in upper and central India up to Peshawar, was complete.

In the first six months of its opening to the public the Indian telegraph sent a total number of 9971 messages of which 8533 were prepaid telegrams on private business. Of this, Indians sent more than a third (see Table 1.1). It was recorded by O'Shaughnessy that while the

> European community are comparatively a very limited class, the native merchants, bankers and fund holders and gentry, may be considered innumerable. The number of native correspondents is accordingly increasing daily. Not only do they use the lines for financial business, but on the very most delicate and secret matters affecting family arrangements, betrothals, marriages, and other domestic affairs, of which they treat with absence of all disguise, which is almost beyond all belief.[54]

Some of the regular business correspondents were firms and merchants of Bombay and Calcutta like Bomanjee Framjee, Gopauldoss, Cursandos, Runchoordoss, Nusserwanjee and Company, Cama and Company, Parmanand Parboodoss, Jootha Dial, Chaneeram Jairaj and Ramjee Madowjee.[55]

Table 1.1 Distribution and percentage of private messages sent by Indians from 1 February to 31 July 1855

Circle	Total Private Messages	Number of Indigenous Messages	Percentage
Calcutta to Agra	4139	1549	37.10%
Agra to Bombay	1735	835	48.12%
Meerut and Punjab	646	153	24%
Bombay to Madras	2013	327	16.24%

Note: Appendix. Abstract statement showing the total cash receipts, and pro-forma charges of each month, on account of Paid and Service messages transmitted by Electric telegraph during the year 1856–7; also the total number of messages sent by 'Natives'. Compiled by Sheeb Chunder Nundee, In Charge, Office of the Officiating Superintendent, Electric Telegraph in India. OIOC.
Source: V/24/ 4282. *Annual Report of the Telegraph Department 1859–1860*

The bulk of the indigenous business was between Calcutta and Bombay (see Table 1.3). The department was surprised at the ease with which Indians used the telegraph for personal matters and admired the communication practices of indigenous business. Commenting on the number of errors in the messages sent, O'Shaughnessy recorded that the 'native merchant never attempts petty economy' and contrasted European complaints with 'the safety and precision with which the vast number of messages sent by native opium speculators, are transmitted between Calcutta and Bombay.' They repeated important figures with words or expressed them twice, for example, Rs. 834 or twice Rs. 417. He argued that 'if the European merchants would adopt this practice, we could transmit their despatches as correctly as any other kind of business we transact.'[56]

O'Shaughnessy noted that the 'important mart' of Mirzapur only returned a monthly average of Rs. 162.[57] This suggests, first, that the Indian business community, especially those engaged in opium speculation, were more than aware of the shortcomings of the telegraph and devised rational strategies in language and expression to avoid error. Second, that the larger indigenous community was willing to trust the Government's promise of confidentiality and privacy. Finally, it is important to note the relative lack of private commercial correspondence in the case of Mirzapur, a large centre of indigenous banking, trade, and finance; perhaps, a rational response and choice reflecting different kinds of trade. Calcutta and Bombay were much more concerned with seaborne trade and opium speculations abroad. Between May 1856 and April 1857 there were 18,628 messages sent by indigenous users out of a total

Table 1.2 Breakdown of private messages from May 1856 to April 1857, including numbers sent by Indians until 1857

Month	Total of Paid (Private) messages	Total number messages	Total number of messages sent by Indians
May 1856	3,865	4,645	798
June 1856	3,151	3,801	1,127
July 1856	2,973	3,616	849
August 1856	3,363	3,870	1,377
September 1856	3,272	3,840	926
October 1856	4,161	4,717	1,550
November 1856	4,286	4,774	1,137
December 1856	5,083	5,800	2,049
January 1857	5,715	6,799	2,390
February 1857	5,093	5,934	2,079
March 1857	5,227	6,348	2,015
April 1857	5,344	6,397	2,331

Note: Appendix; 'The Total Number of Messages Sent by Natives'. Compiled by Sheeb Chunder Nundee, In Charge of Office of the Officiating Superintendent, Telegraphs in India. OIOC. One of the possible explanations behind the seasonal fluctuations between May 1856 and November 1856 is that the monsoon season frequently put the telegraphs out of order. Another explanation would be that crops including commercial crops like opium are harvested during May, July, and September.
Source: V/24/4282. *Annual Report of the Telegraph Department 1859–1860*

Table 1.3 Regional breakdown of telegrams between November 1856 and January 1857, including those sent by Indians

Lines	Total number of messages	Number of Indigenous messages
Calcutta-Agra line	2170	282
Calcutta-Bombay line	3321	814
Bombay-Madras line	257	72
Madras line	745	186
Meerut-Punjab line	385	65

Source: Home Public Proceedings, Electric Telegraph, no.1, 20 March 1857, from Lieutenant P. Stewart, Officiating Superintendent, to C. Beadon, Secretary, Home Department, Government of India, no. 228, 14 February 1857. NAI

of 60,541 messages worth Rs. 3,10,384 in round figures (see Table 1.2).[58] The increases in June and August reflected the lowering of the tariff for commercial messages by about 25 per cent.

Diffusing and disseminating electrical science and telegraph technology

Europeans and Indians made some significant attempts to diffuse and disseminate knowledge of the railway and the telegraph in the public sphere before 1857. Kalidas Maitra from the town of Srirampur published the first Bengali book on telegraphy in 1855.[59] He campaigned for the establishment of public libraries, for example, in Krishnanagar,[60] composed plays such as *Muktavali*, among others,[61] wrote original works on the telegraph and railway,[62] and published on the principles of anatomy and physiology.[63] He published the bulk of his diverse writings before 1857. Maitra was a conservative social reformer and a critic of Iswar Chandra Vidyasagar's widow-remarriage movement.[64]

After the usual *shastric* citations, Maitra forwarded a simple argument, which is worth noting in order to place his work in an intellectual and social context. First, he argued that Vidyasagar was allowing the British to intervene in the mores of Hindu domestic life. He believed that British laws and government should be kept from intervening in private domains more appropriate for reform from within. Second, he claimed that though there might be a few widows forced into prostitution or in extreme neglect, these were the exceptions. The larger Hindu joint family protected the majority of Hindu widows who played important roles within it. Finally, Maitra suggested Vidyasagar was importing foreign models of social reform and applying them to indigenous society. These would be evangelical models that based themselves on the experience of fragmentary industrialisation in Britain with its attendant exploitation of mostly urban, landless labour of both sexes.

Maitra's work often combined a visionary modernity with conservatism and early nineteenth century nationalist motifs. For example, he attempted to prove that ancient Indians used steam power in *Vaspiyakal o Bharatiya Railway*.[65] Apparently, Raja Shallya, in the *Bhagavadgita*, goes to the master-mechanic, Moydanab, who gives him the *Souvyantra* to destroy the *Yadu kula* (Lord Krishna's clan). This contraption could travel in water, air, and, land, and had a tail of smoke. Obviously, it was a steam-powered machine, and he quoted the relevant *shloka* in proof, of which the last two lines are: *Kacchidbhoumo kacchidyomni girimuthi jale kacchit, Alatchakrabat bhramyat soubhong taddurabsthitang* (Sometimes on land sometimes in air, on tops of mountains, in water sometimes, like a circle of fire, travelling, the *soubhya* rests there). (Translation mine.)[66]

According to him, there were other such examples in many texts including the *Puranas*. Similarly, he explained the inclusion of the list of the rulers of *Bharatbarsha* or India from Yudisther, the mythological figure from *Mahabharata*, down to the Victorian present by claiming correspondence between events in the Bible and the *Puranas*. He states that reading his book, the reader would realise the futility of hate and contempt between Indians and Europeans.[67] Maitra perceptively explained the British state's strategic need to rapidly transport troops lay behind the establishment of the railway in British India.[68] He admitted taking his sections on steam, what constituted a steam engine, and its modern history, from English books.[69] He asked the rhetorical question, 'What is steam?', revealing a laudatory ability to get to the basics of his subject.[70] Maitra was an unabashed champion of machines of modernity. He saw the space and time contraction enabled by the railway and telegraphy as exemplars of modernity and civilisation, and urged his compatriots to learn from the European attitude that time was money. His was a curious combination of early nationalist egotism and conservative radicalism.

Maitra, in his book on telegraphy, claimed that though authors like M. Townshend, J. Robinson and Reverend J. Mack published articles on electricity and chemistry in *Tattobodhini*, *Bibidhartha Sangraha* and *Satyapradip patrikas*, his book was the first in Bengali or Sanskrit to cover the whole subject of electricity.[71] Starting from the first principles of magnetism and chemical reaction he provided guides to experiments that could be conducted using household goods. He called telegraph stations *addas* or, ironically in today's world, 'chatting spaces' and described O'Shaughnessy's Single Needle signalling instrument in detail over 11 pages, perhaps in the hopes of Indian duplication.[72]

Maitra revealed a fundamental truth in his book: the dots-and-dashes sent by telegraph had no necessary connection with what they signified.[73] He wrote, '*Sanket aicchik janiben ... atabata engraji akshare jeiroop sambad asiya thake tadroop bangala ki parasya prabhriti bhashay sambad asite pare*' ('Know that signals are arbitrary ... therefore similar to how information arrives in the English alphabet, news can come in Bengali or Persian or other languages' [translation mine]).[74] Both the signifier and the signified co-existed in arbitrary though self-consistent systems, that is, dots-and-dashes existed with their system of differences and could be used to convey practically any information in any language based on a system of internal differentiation. Maitra provided the skeleton outline of systems of signalling Hindustani, Persian, and Bengali alphabets alongside the English one.[75]

Similarly, Akshoy Kumar Dutta wrote a Bengali manual for those wishing to travel by train. Clearly identifying increasing speed and facility of transport and communication with modernity, indeed *sabhyata* or civilization, Dutta expressed his optimism for the future development of India (*Bharatbarsha*) but warned that nothing on earth was an unblemished good.[76] Compiled from English guides, this book not only gave rates and distances, especially rates charged for bulk goods, but a significant portion of it was devoted to the do's and don'ts of railway travelling. For example, do not stick head, feet, and other parts of the body out of the window; do not jump off a moving train; or jump off a train that has stopped because a train coming behind will crash into it, but remember to jump on the side that is free of rail tracks, etc. What is more fascinating is that this guide laid down directions for how to live with the railways running through one's area, for example, when to cross tracks, where to sit or stand near the track, how to know whether a train is coming around a bend, etc. These early attempts to disseminate technology became irrelevant after 1857 as technology became centred in Britain and racially exclusive, Maitra saw a vision of a country, which was not mediated solely and exclusively by a foreign language. In contrast, China adopted a number system for Mandarin and used its own language in telegraphy. By 1911, the Qing had nationalised the telegraphs. This possibly had a fundamental differential impact on national cohesion of the two countries and the cost-effectiveness of telegraphy because messages had to be first translated into English and then re-translated back to the vernacular of the recipient. Increasingly, the railways and telegraphs became objects of satire, sites of pollution, and symbols of an oppressive modernity.[77]

Conclusion

The period 1830–50 was one of fundamental change for the colonial state. Over this period it evolved the basic template for its attitudes and policies towards the country it had come to rule over. Science and technology became institutionalised, the Asiatic Society took up orientalist research, and the telegraph language was exclusively in English. These changes had profound consequences for the future of India. A telegram became doubly expensive since drafters and translators had to receive their profits. The policy of using English created new elites and rendered older ones redundant. More importantly, it created schisms and divides between Indians. The transition from individualistic experimentation to mass implementation meant that India received very few spin-off effects and had little chance of substituting technology. Early attempts to translate and diffuse knowledge of steam, railways, electricity, chemistry, and telegraphy also became anachronistic as the colonial state created exclusive racial preserves, restricting technology and shrouding it in secrecy. Secrecy dominated the moral economy of scientific enterprise in British India while increasingly the bureaucratic principle dominated its structure. The use of telegraphs by Indians did not mean they understood how telegraphs worked. After 1857 there was no chance that knowledge of the technology would be diffused in larger society, or that the use of Indian languages would allow the technology to be a simple 'national legacy', a neutral technology flowing seamlessly from the colonial to the national.

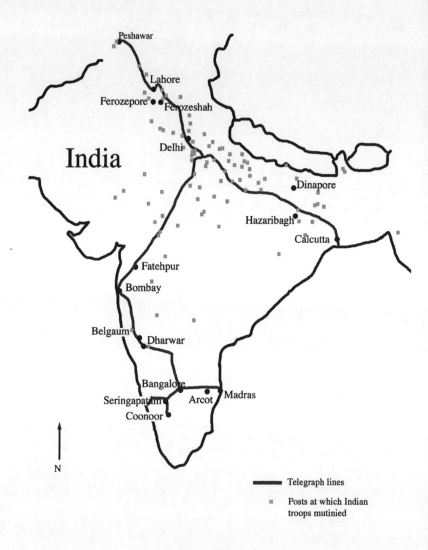

Map 2.1 Uprisings of '1857' and the telegraph lines

2
The Telegraph and the Uprisings of 1857

Introduction

The uprisings of 1857 were a major turning point in the history of British rule in India and an equally major test for the telegraph system recently built in India. 1857 was a communication crisis of enormous proportions for the British in India, and the telegraph has conventionally played a redemptive role in the huge volume of narratives generated around the uprisings. These uprisings cannot be limited to the years 1857 and 1858, and they were still echoing in central India while the professional and English speaking literati were negotiating the founding of the Indian National Congress in the port cities in 1885. So the term '1857' has a much wider meaning both in terms of content and extent than the evocation of a single year indicates: a similar example is the term '1789' in the history of France (though Stokes preferred the parallel with 1848: a 'failed revolution').[1]

Telegrams sent by the signallers at Delhi warning of the rebellion were crucial to the British retaining their control over the Punjab, thus saving British power in India. 'The telegraph saved us', claimed Sir John Lawrence of the Punjab. The original statement by Montgomery, Judicial Commissioner of the Punjab, was, 'The electric telegraph has saved India', and it was he who received the telegram sent by William Brendish, Signaller, warning of the emeute at Meerut. Similarly, an anonymous sepoy, waiting to be hanged, pointing to the overhead telegraph wire, allegedly exclaimed, 'that is the accursed wire that strangled us.'[2] It is these two opposed voices according the same importance to the telegraph that is the focus of this chapter.

Though the telegraph signalled the warning that saved the Punjab and diverted troops by sea to Calcutta, this chapter argues that the myth of a successful telegraph was fundamental to the propaganda of 'armed science' and the idea of an empire grounded in science and technology. In reality, there were fundamental defects in the construction and spatial organisation of the telegraph in India that jeopardised it, independent of the uprisings. This chapter examines the issues involved in the context of the information strategies deployed during the events to argue that many of the decisions of the government concerning information control reflected its fear of a 'white mutiny' and the need to restrain its own information leakages. The relationship between technological failure and the mutiny-related information panic suggests that systemic crises occur almost simultaneously within technology and society. 1857 was a particular conjuncture of different factors and movements that included the technological, yet this chapter also shows how the larger information crisis subsumed and reinterpreted localised technological failure.

Technological flaws: Design and insulation

There were serious technical flaws in the system. Defects in design were attributed to W. B. O'Shaughnessy's pride in claiming to be the pioneer of 'Indian' telegraphy. The first problem was non-insulation and many of the Indian telegraph lines were not insulated. Subsequently, he used his influence in England to patent the 'Brook Insulator',[3] which was a vulcanite cap slotted into an iron hood with sulphur cement that proved useless in practice. Ceramic insulators were established to be the best. These together damaged the system irrespective of the uprisings, and the uprisings were a boon in disguise for the department, which used reconstruction as an excuse for extensive modification in design and structure. In 1866, the Director General of the telegraph attributed the bad working of the early lines to their being 'inflicted' with the Brook bracket and insulator, 'both of which are thoroughly unfit for the purpose for which they were designed.'[4]

O'Shaughnessy's obsession with his own method of insulation had other difficulties. Initially, he claimed to have dispensed with the need for insulation through the use of iron rods weighing as much as a ton to the mile, and Luke, writing in the 1890s, noted the existence in India of thousands of miles of circuit weighing half a ton to a mile.[5] Though he then patented his own brand of insulation, blatantly contradicting his previous claim of the feasibility of non-insulation, he continued to prefer extremely heavy iron rods rather than wires. The problem with this was that the quality of the transmission of signals was very poor.

Messages were often swallowed up at each successive station, that is, the original message regularly lost its final sections until it became indecipherable or incomprehensible at best, or non-existent at worst.⁶ In his report of 1862, O'Shaughnessy confessed 'The Indian lines offer the only examples in the world of lines entirely devoid of insulation ... the result of automatic transmission through eighteen successive instruments, as on the line to Bombay, would not merely render the message unintelligible, but in extreme cases, the message would, during transmission, entirely disappear.'⁷

In August 1857, during the height of the mutinies, the line between Raniganj and Benares was inoperable for 14 days because of faulty insulation. This was a strategically crucial distance of 300 miles connecting the last railway station with the first military cantonment on the river Ganga. Harriet Tytler in her account of 1857 revealed the public doubt about the efficacy of the telegraph. She wrote upon meeting the mail cart from Delhi that

> [t]his mail cart leaving Delhi punctually at the appointed time with the mails as if nothing had happened showed the subtlety of the ... Asiatic character. They knew the wires being cut would cause no surprise to the authorities elsewhere, because telegraphy was in those days in its infancy in India and messages were often delayed through something going wrong with the wires, but the post not arriving at the appointed time would certainly cause suspicion and enquiry.⁸

Certainly O'Shaughnessy antagonised many people: agents of British patent holding companies, railway engineers and some of his colleagues. The problem lay between his desire to innovate and create original systems and the government's need to regularise and standardise communication systems. O'Shaughnessy failed to serve both these ends. Even discounting the criticism of Adley, an advocate of privatisation, free trade and railway companies, and his successor as Director General of the Telegraph Department, D. G. Robinson, there were several defects in the telegraph system of 1856.

The original instruments designed by O'Shaughnessy were still in use during 1857. They proved unreliable both in terms of accuracy and the method employed to translate signals. An increasing volume of public traffic did not afford him the luxury of further experimentation. A small number of imported Morse instruments were introduced into 1856 in India. The Bangalore workshop produced a number of substitutes, but these later proved unreliable.⁹ For telegraph signallers recruited hastily from the military and from different regions, this lack of uniformity of practice and instrumentation presented severe operational difficulties during 1857.

Structural flaws of the Indian telegraph: The choice of routes

The extremely linear routes adopted by the telegraph were particularly vulnerable to disruption by their very nature. Lord Dalhousie chose the shortest and quickest route to 'meet the immediate necessities of the time.'[10] By 1856, a line was built from Calcutta to Peshawar; Agra, Head Quarters of North West Province, was joined to Bombay; and, Bombay linked to Madras. There were no alternative routes and in their haste the government ignored local conditions to geometrically construct the lines. As Edney points out, '... the rational, uniform space of the British maps of India was not a neutral, value-free space. Rather, it was a space imbued with power relations ... the British suborned the geographical character of those territories to a mathematical space ... an exercise in discipline.'[11] This disciplining of space was never so obvious than in the telegraph network inaugurated in 1856. It was a linear network that linked the commercial and military nodes of the colonial state and had little to do with indigenous routes of trade and communication in the subcontinent.[12] The East India Company in the past had often moved away from established centres of communication and commerce to avoid tolls, taxes, strong middlemen and entrenched interest groups that focused around these centres.

J. Deloche argues for the natural or traditional structure of communications in a country: an organic order. In the case of India he identifies an inverted 'Z' manifest in the shape of traditional communication routes in the subcontinent.[13] According to Deloche, the main axis ran from the North West through the Ganga-Yamuna Doab to the Ganga-Brahmaputra delta; Kabul (the route continued into Europe), Peshawar, Delhi, Benares, Rangpur and, south to the Ganga-Brahmaputra delta or continuing through Cooch Behar into Assam, and beyond. The second axis connected the northern plains to the Gulf of Khambat. The third axis crossed the Deccan to the Coromandel Coast while the last loop crossed the Pal Ghat Gap to the western coast. The Mughal system with its central node in Delhi had a main central line, which serviced Kabul, Lahore, Multan, Agra, Allahabad, Benares and Patna. Surat and the Gulf of Khambat were the convergence of two lines from central India. The first passed through Rajputana via Ajmer and Ahmedabad. The British returned to this route during 1857 to connect Agra and the Punjab with Bombay. The second passed through Malwa via Burhanpur. The connection between Surat and the ports on the eastern side of the Deccan plateau was serviced through two centres, Aurangabad and Hyderabad. Deloche observes the opening up of the Gondwana to late Mughal and Maratha State communications.

When Lord Dalhousie called the establishment of the telegraph in India a 'national experiment', the nation he was referring to was not India.[14] What it came to mean was a national investment for Britain and an experiment in empire. Britain was to be the laboratory and India the empty space for the enactment of the telegraph experiment. The 'public' of the colonial state were the people of India but the nation to it was the metropolis. This disjunction between public and national is important in understanding the nature of the telegraph system first built in India and its subsequent disarray in 1857. There were several routes proposed for the telegraph system. Charles Adley proposed that Calcutta, Peshawar, Agra, and Bombay should be joined with another loop connecting Agra, Bombay, Madras, and Calcutta. This would have allowed a fall back system to the main one. While the main line to Peshawar was complete by 1856, the inner loop connecting the three colonial metropolises with Awadh and Agra was not started.[15]

The government vetoed the proposal because it was perceived to be of secondary importance to the main line to Peshawar via Agra. This dangerously isolated Agra and the North Western Province during the uprisings of 1857. The government of Bengal suggested that the Ganga River be followed until Benares. Deviating from this suggestion, the telegraph line travelled south and traversed the northern reaches of the Kaimur plateau to turn north to Sherghati, a little below Gaya, which is on the Ganga. Sherghati had little claim to importance except that it was easiest to cross the Son River at that point. The line then proceeded past Chunar across the Ganga to Benares. The Mughal route followed the river almost all of the way. The Calcutta-Sherghati route followed a much more southern direction, meeting the river above the Kaimur plateau: this was a continuation from the late Mughal penetration of the Gondwana region.

The East India Company's administration, by the end of the eighteenth century, had, at least twice, shifted the Grand Trunk Road away from the traditional central axis that had followed the Ganga River. Similar shifts occurred in Bombay and Madras Presidency. For example, in the late eighteenth century a route was forged through Indore to Agra at the cost of the older Ujjain to Mathura road; a major indigenous banking route.

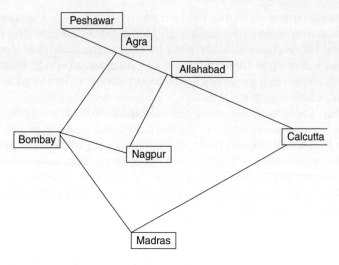

Figure 2.1 Proposed structure of telegraph lines

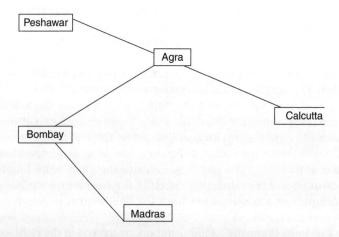

Figure 2.2 The telegraph system that was completed in 1856
Note: The drawings are not to scale and represent the structure of lines rather than actual distances or angles.
Source: See, letter from Charles Adley to Sir Archibald Galloway, Chairman of the Court of Directors of the East India Company, September 1850; proposing that the 'best plan to be adopted would be that known by the term "circular system"', Adley, *The Story of the Telegraph*, p. 8.

The telegraph line connecting Calcutta with Madras was not built. This presented a problem in 1857 and Calcutta was starved of official information because of the breakdown of the government's communication network. At the height of 1857, Agra was strategically isolated until a camel *dak* was started through Rajputana, passing Deesa, Ajmer or Nasirabad, Jaipur and Bharatpur to Agra. This was a return to the traditional axis between the western coast and northern India. Another route adopted was the route below Jind and Patiala onto Lahore and Peshawar. The main road, joining Bombay to Agra was in the hands of the rebels during the uprisings. Running through central India, this road passed through Mau [Mhow], Indore, Dewas, Sipri and Gwalior. Nearly all of these centres were significant nodes of the 1857 uprising, and so when the disruptions occurred, the government had no alternative routes or supports to ensure the continuation of telegraphic communication in central India.[16]

Figures 2.1 and 2.2 illustrate the difference between the ideal system suggested by many senior officials and what was actually constructed. The lack of a link between Calcutta and Madras and the absence of circuits or duplicate lines rendered the system inherently vulnerable. After 1857 the principle of alternate and duplicate lines was adopted in every case. The submarine and land systems would never again be so linear or rational, and indigenous demand was catered to after 1857, in part to provide for alternate routes if a single route failed. It was admitted that on the first establishment of the telegraph it was 'considered desirable, in view to a saving of distance, to carry the lines as nearly as possible direct', thus passing through impassable country remote from roads and public spaces.[17]

Strategies of censorship: The need to contain information

The main problem facing the Company's Government at the onset of the uprisings was the containment of information that was considered inflammatory. As rumours of routs and massacres multiplied in Britain and India, the government was hard pressed to control panics. As early as 1856 the containment of information had become a serious problem. Dalhousie's precarious policy of refusing recognition to adoption, illegitimate sons, and minors, and extension of control over large parts of central India led to rulers, chieftains and talukdars becoming anxious about their future. For example, James Blacknight, in charge of the Bombay Telegraph Office, precipitated a political crisis by leaking an important confidential telegram from the Government of India in Calcutta sent through the Bombay office to the Political Residents at Pune and Nasik. The wording in the telegram that was leaked by Blacknight was amazingly summary: 'Tell Major Davidson to acknowledge no successor to the Gaekwar, to form a council of administration to carry on the Government ... Major Davidson must carry out these instructions in such a manner as to attract as little attention as possible.'[18]

The *Bombay Telegraph Courier* published a premature editorial proclaiming:

> We can now confidently announce that the last of the Gaekwars has been gathered to his fathers. It is expected that this step [annexation of Baroda] will excite universal surprise and consternation in Baroda, but it is anticipated that no emeute will take place in consequence. ... It is the duty of the Empire to absorb all those petty states, where bloated human vampires commit deeds of cruelty and oppression ... a leading article on the annexation of Baroda in the *Times* or the *Daily News* will dissipate all the froth and vapour, which Baroda gold may purchase to disturb and annoy the British Parliament when it assembles. With Oude sauce and other Indian condiments, ministers are likely to have some severe labour during the ensuing season.[19]

This quotation illustrates the extent to which both sides, Indian and British, manipulated public and political opinion in Britain and India.[20] It also reveals the extent to which the Company's policy as well as the public perception of it had ingrained within it the principle of direct and indirect annexation. Leakages like this convinced other rulers that their kingdoms were going to be taken over as a matter of course.

James Blacknight was an interesting case because during the first six months of the telegraph, O'Shaughnessy recorded in his report on the establishment, *'cannot read signals*, Head Assistant, character good, a very intelligent, well informed, and useful office man.'[21] In his defence, Blacknight pointed out:

> The Bombay signal office is even now the great centre through which passes most of the important despatches from Bengal and the North West Frontier to Bombay and Madras and vice versa. This is also the great centre of opium speculation, and here, where secrecy is essential the duties have been carried on in the presence of twelve to fourteen hands on duty at a time, without the slightest possibility of secrecy.

He suggested that the authorities 'build or create' some suitable building where 'secrecy can be endured'.²² Blacknight thereafter fled to Britain with the cash in the office. His parting telegram stated that he would meet O'Shaughnessy in London to present his case.²³ The government was not amused, and along with extradition proceedings, requested that the officiating Director General report on the measures adopted to preserve the secrecy of important messages pending the introduction of a cipher being prepared by O'Shaughnessy.²⁴ The outbreak was upon the department before these procedures of containment were set in motion yet these concerns reflected the obsession of a ruling power with control over information and the containment of information. Instances of information leakage increased by the middle of 1857 and the telegraph was repeatedly cited as very vulnerable to such leakage.²⁵

The East India Company's Government in Calcutta gagged the press after the outbreak of 1857 to not only control the relatively docile vernacular press but to curb the strident and panic-prone Anglo-Indian press. The indigenous press in both the Bengal and Bombay Presidencies remained largely loyalist in their tone. A series of pamphlets in Gujarati and Marathi, written by the loyalist indigenous editor of the *Bombay Times*, were circulated in the Bombay Presidency. These compared the conditions of life under the Mughal, Maratha, Gujarati Hindu and the Company's Government to extol the virtues of British rule.²⁶ In contrast, in Calcutta the Anglo-Indian *Harkaru* paper was suppressed because of a series of letters from a person calling himself 'Militaire'. In a letter called 'Our future rule in India', symptomatic of the gratuitous and alarmist expatriate propaganda of the period, the author wrote:

> Now is the time to show the sepoys we can do without them ... this can only be done by bringing up Africans from the Cape of Good Hope ... and before next hot weather we shall have a body of men who can stand the sun as well as the natives of India, and who, with the Sikhs and Gurkhas, will relieve our European soldier from all the common duties which involve exposure to the sun. ... But no time should be lost in beginning to form an army to replace the Brahmins and treacherous Mussalmans.²⁷

Such advice from expatriates, amounting to criticism of the government and the implicit claim to be a part of government, was obnoxious to the increasingly rigidly hierarchical officialdom of the Government of India. The government followed a similar policy in the Punjab, censorship and disarming of the native regiments. Fears of a 'white mutiny', such as occurred in Madras, and the shade of the American Revolution haunted the East India Company's Government, perhaps not without reason. There was a Kashmiri/European brothel madam, who was executed after the British returned to control, taunting the sepoys at Meerut for being impotent in not saving their fellow soldiers from dishonour, possibly leading to the start of the mutiny. There is also the example of someone called Schmidt who tried to convince the civilians and non-commissioned elements in the garrison to rise against their superiors.[28]

Destruction of the telegraph and the crisis of communication

Over 918 miles of telegraph wire were destroyed in the uprising. The scale and ferocity of the destruction of the telegraph lines suggests that it was a primary military target for the rebels. If one examines the map, it appears that much of the military action occurred along telegraph nodes and routes. This was in stark contrast to the 1855-6 Santhal uprisings that left the wires unscathed. If the telegram that saved the Punjab had not been sent, the conflagration would have been much worse because the sepoy strategy, as the map shows, was to destroy the telegraph since they lacked the knowledge to use it. That the rebels were very aware of the telegraph is proved by the rapidity with which they severed lines and concentrated on telegraph nodes such as Sherghati in Bihar or Mau near Indore. Running out of ammunition, rebel soldiers used the wires to fashion crude bullets and the posts were chopped for firewood. The recently introduced wrought iron tubular telegraph poles were perforated and used as cannons by the sepoys. O'Shaughnessy, on his return, found the lines in the North West Province totally destroyed as were the lines from Agra to the Narmada, and signallers 'murdered or dispersed, and that the Madras and Bombay lines had, from want of supervision, fallen into total disorganisation'.[29] In January 1858 he had less than 2500 miles of functioning wire and only 50 offices in efficient operation.[30] The line between Meerut and Delhi was the first to be destroyed. The British subsequently opened a telegraph office at Aligarh, an old *quasbah*, but after ten days they retreated to Hatras.[31] Between June and September 1857 the British fell back on traditional means of communication, routes, and of sending information, which their reconstruction of indigenous space had ignored since the 1830s.[32] What emerges from the narratives of 1857 is the continuous battle over communications.

The major problem facing the Government in Calcutta was the near complete isolation of Agra, headquarters of the recently formed North West Province. Agra was isolated from the first week of June.[33] On the first of July, the Indore outbreak took place leading to the deaths of Butler, Avery, Bone, and Brooke of the Telegraph Department; the offices, lines, and instruments were completely destroyed.[34] This was the last line that had kept telegraphic communication open between Agra and Bombay. By July 1857, 400 miles from Agra to Indore, 180 miles from Kanpur to Agra, and 178 miles from Agra to Delhi had been totally destroyed. Between Allahabad and Delhi there was no telegraphic communication, the Narmada River region was overrun, and between Kanpur and Allahabad there was nothing.[35] Charles Todd, Assistant Telegraph Inspector in charge at Delhi, was killed. The Kanpur establishment was wiped out. Between Allahabad and Benares, where the merchants were relatively loyal, only a small amount of damage was done and though the wire was cut several times, two employees of the Telegraph Department, Devere and Babu Thakur Prashad, managed to restore communications.[36]

By September a camel dak, through the Kutch and Rajasthan, secured communication between the Punjab and Bombay.[37] Telegraphers sometimes resorted to burying the telegraph instruments to prevent them being destroyed by the opposing forces. In Alam Bagh in Awadh a temporary semaphore was built and the knowledge to do so came providentially from the *Penny Encyclopaedia*.[38] During this period the Post Office, in charge of the Bullock Train and Tonga dak, transported over 18,000 troops up country from Raniganj. In contrast, after the monsoon, river transport managed to carry around 5800 troops. The postal workers scattered throughout the country were valuable sources of intelligence about the movements of the rebel forces.[39] Regular mail was disrupted between Agra and Bombay until February 1858: a total of seven months and thirteen days.[40] Captain J. G. Medley, writing from the Punjab, noted in July 1857, 'Beyond Delhi our knowledge was a blank. The whole country was in the enemy's hands and our only means of communication was round by Bombay and Calcutta, where the ignorance of what was passing between Allahabad and Delhi was as great as our own.'[41]

It was a battle over communications and the telegraph situation was like a tidal pattern with new offices continuously being opened, destroyed and reopened in different centres. In the absence of the telegraph, letters were sent on tiny slips of parchment or leather that were secreted on the *quassid*. They were written in Greek, Latin and French while the other side used Hindustani, Sanskrit, Urdu and Persian. Both sides used codes and were prompt to execute couriers and informants.[42]

It was a battle between alternate routes and networks with merchants supporting both sides as they allowed their networks of *hundi* and special couriers to be used. In many ways, 1857 was as much a contest of local knowledge, as it deployed modern strategies of disruption, counter-propaganda and publicity.[43] Such wealth of detail is an essential quality of the narratives of 1857, reflecting the turn to local knowledge and individual experience.

The British, in 1857, did not have much archival knowledge or control over central India in terms of maps and surveys. When Major General Sir Henry Havelock marched with his troops for the relief of Lucknow he had no adequate map to rely upon. In Calcutta his search had produced 'nothing except a rough plan of the high road to Lucknow, sketched ten years before, which was not only imperfect, but so inaccurate, as to be worse than useless for military purposes.'[44] An engineer from the Oudh Railway Company had scientifically surveyed the line from Kanpur to Lucknow only four months before the outbreak, but he died with General Wheeler's force and all his papers and plans were destroyed in the sack of Kanpur. Between Kanpur and Lucknow no trace of the telegraph remained when Lieutenant Patrick Stewart went over the road with General Havelock's column. Havelock relied on local informants and scouts to know the land. Telegraph technology, like scientific surveying and institutionalised information, failed in several fundamental ways to be of much use in the crisis.

Interception of communication

Obsession with secrecy and the interception of communication were important features of the uprisings of '1857'. Both sides employed traditional as well as modern methods of communication and propaganda.[45] By October 1857, the government prevented magistrates from transmitting any telegram that gave news of government losses before it was officially published. Signallers were to be imprisoned and fined Rs. 200 for any disclosure of official messages. The government proposed that private messages should be countersigned by a high official before they were transmitted.[46] Perhaps, it is important at this juncture to examine the volume of messages sent over the telegraph. The record until November 1856 shows that the number of native messages sent was 1137. The proportion of indigenous users was small and probably concentrated around opium speculation, that is, speculation on prices at the port cities and production information from the hinterland so the bulk of the information carried was over long distances.

The Post Office still carried the bulk of the mail of the country and it is there that the British Government practiced the most systematic interception. Letters found in the Dead Letter office in Madras confirmed the suspicion that the post was being used for 'treasonable purposes' and all magistrates were authorised to open indigenous correspondence with discretion. Plots began to be unearthed as far way as Chingleput and Coimbatore while a seditionist was imprisoned in Tanjore.[47] The sole object of placing the Post Office under the supervision of the civil authorities was to:

> [P]rotect the public and private correspondence of the country being tampered with by any native Deputy Post Masters, who might be in communication with the rebels ... a covenanted officer should be in a position to know whether native letters, especially letters from the North West, in Persian or in Hindi, are passing through the district ... whether any printed pamphlets or native newspapers, especially Persian or Urdu, are increasing in circulation; and whether, in short there are any communications of any sort, generating suspicion, and therefore requiring to be checked.[48]

However the Judge clarified that the Bengali Post Master at Jessore was unlikely to be in 'correspondence with rebels, who have beaten and even slaughtered his countrymen.'[49] Two seditious letters, one in Sanskrit from the 'notorious' Muniram, hiding in Calcutta and addressed to a pundit in a government school signed in code with the key given in verse, and the other written by an Assamese, but without signature or date, were intercepted. Along with orders to arrest Munshi Muniram of Assam, orders were issued to arrest the dethroned Saring Raja on charges of sedition.[50] This illustrates both the extent to which an India existed across Indian cultures and how quickly technologies such as printing and communication systems were deployed in indigenous organisation. The diffusion of knowledge about telegraphy needs to be distinguished from diffusion of use.

Codes and ciphers

Sir W. B. O'Shaughnessy returned to India to assume charge of the Telegraph Department in November 1857.[51] He designed a code for the telegraph while in England, copies of which were forwarded to select senior officials in September. It was above all designed to keep signallers from being privy to government secrets. The recipients of the code were the Secretary, Home Department; Secretary, Foreign Department; Lieutenant Governor, Central Provinces; Chief Commissioner, Awadh; Private Secretary to the Governor General. The government warned

> utmost precaution must be taken to prevent the cipher being used or seen by any but confidential persons employed for the purpose preparing or deciphering messages. On no account is the cipher to be communicated or made known to the officers of the telegraph Department. It is intended, and it is necessary, that contents of messages transmitted in cipher should not be understood by the signallers.[52]

The government clarified that the code was not meant for all government messages[53] but only for 'such messages or parts of messages, as it is of real importance to keep secret' and that copies of the code should not be available to a greater number of people than was absolutely necessary.[54] Thus, in 1857 the government not only resorted to the suppression of the press but also deskilled the signallers so that they could pose no threat to government control over information.

Propaganda, counter-propaganda, and rumour: *Sandhyabhasha*[55]

Secrecy and symbol were important in 1857 and both sides used propaganda extensively. As a result, the distinction between rumour and news proved difficult to determine. The *Delhi Gazette* complained 'so many reports and scraps of intelligence are daily brought to us, and some so directly contradictory of others, that in our desire to publish nothing but facts authenticated or nearly so, we are often at a loss what to place before our readers.'[56] Soon after the passing of the Act xv of 1857 imposing censorship over the press, Robert Knight, editor of the *Bombay Times*, asked the government after it had censured the newspaper for publishing two articles relying on baseless rumours widely held to be true at the time of publication, that 'some means be afforded the Press of ascertaining the truth or falsity of rumours.'[57] The government in reply instructed that 'whenever you receive a report of a mutiny you should communicate with the Secretary to the Political and Judicial Department, who will inform you whether any *official intelligence* of the event has been received, and if official intelligence has not been received, the report should not be published' (italics mine).[58]

Indigenous sources continued to provide quicker news than the official apparatus. For example, they reported the victory of the British at

Bulandshahr much before official confirmation arrived.[59] The world of 'bazaar gossip and rumour' that appeared incomprehensible to the British repeatedly proved more effective in information gathering and dissemination than the state. Equally, if a telegram bearing news of the war did manage to escape censorship there was no way of knowing whether the information was any more reliable or swift than any other network. As Robert Knight's predicament illustrates, rumours were very much the staple of information networks in 1857. If the East India Company's Government resorted to censorship and codes to prevent information leaks, the rebels also employed counter codes and propaganda.

The strength and vitality of other modes of transmission and communication emerged vividly in the course of the uprisings. During the spring of 1857 rumour seemed to increase in virulence and numbers. Sir George Trevelyan recorded that the 'native society of Hindostan presented that remarkable phenomena which, in an Asiatic community, are the infallible symptoms of an approaching convulsion. ... No one can tell whence the dim whisper first arose, or what it may portend; *for the Hindoo like the Greeks of ancient time, hold Rumour to be divine*' (emphasis mine).[60] Equally, like codes, prophecies and rumours abounded in 1857 and both the British and the rebels listened earnestly to them. One prophecy that had wide currency at the time was the predictions of Neamatullah Shah, whose mausoleum was in Kashmir. It seemed to be uncannily precise:

> When that [Muslim] King is dead and gone, in his house a fracture will take place
> [And] the Clan of Sikhs will exercise over the Mahomedans great tyranny and oppression; for forty years this great tyranny and heresy will remain.
> After this the Nazarene will seize the whole empire of Hindustan
> [And] for the space of one hundred years, their sovereignty will remain in Hindustan.
> When in their time heresy and tyranny shall become general
> For their assassination Shah Ghurbee [Western King] shall appear
> [And] between the two will be fought desperate battles.
> By the strength of the Crescentader's sword [*beyjor-e-tej/taj-e-jihad*] the King of the West will be victorious.
> [Then] Without doubt the followers of the Clan of Jesus will be broken and dispersed...[61]

The printed copy of this prophecy was published in Delhi by the press of Syed Jamalluddin and its extensive circulation was achieved by slipping it in as a supplement to the large number of vernacular papers published at that city. It is difficult to put an exact date to it, but it was certainly circulating before 1857.[62] The year 1857 saw the completion of the first 100 years of the onset of British rule in India in Bengal, after the Battle of Plassey in 1757. To the contemporary official, the prophecy probably appeared to predict the intervention of the Amir of Afghanistan. The British Government had signed a treaty recently with the Amir and the Government of India mistrusted him. The links between Calcutta, Delhi, Peshawar and Kabul were still very strong and the memories of Nadir Shah and Ahmad Shah Abdali still fresh. Some of the rebels and leaders hoped that the Amir of Afghanistan would intervene to save their cause.

The famous passing of the *chapatis* is a classic example of an information panic. Trevelyan wrote, 'During the early days of March, every hamlet in the Gangetic provinces received from its neighbour two chupatties, the staple food of the population ... mysterious symbol that flew, and spread through the length and breadth of the land confusion and questioning, a wild terror, and a wilder hope.'[63] The extremist nationalist, Savarkar wrote that the *chapatis* 'set the mind of the whole country on fire by the very vagueness of the message.'[64] Both authors possibly exaggerated the extent to which the *chapatis* circulated and some later historians have indicated reasons other than the mutiny, such as the potential for a break out of an epidemic disease such as smallpox, which might have caused their circulation. The point is that rumour and news are not easily separable and rumour as prediction often constituted news in this period. However, the reading of such symbols became a British obsession after 1857. The excess and wide variety of meaning that could be imputed to such mute symbols were a nightmare for them, though they meticulously recorded and tried to interpret the passing of similar signs and symbols throughout their rule; perhaps, ironically, the British created their own breed of astrologers, readers of signs and portents, prophesiers of the future, detectors of treason and sedition.

Neither technology nor divine intervention proved to be saviours ultimately. Goddess Kali was invoked and it was rumoured that she would give invincibility from bullets. The following verse was inscribed on the robes of Shankar Sahai, rebel Raja of Jabbalpur:

> O Mahakali! Cut up the backbiters; trample under thy feet the wicked; ground down the enemies, the British, to dust. ... Destroy their common servants and children. ... Protect Shankar Sahai.

His ancestors had occupied the *gaddi* or seat of Gond for 1500 years until the Marathas overthrew them. The company's troops defeated the uprising led by Shankar Sahai. Shankar Sahai was captured and was tied to a cannon's mouth and blown up.[65] Prophetic and supernatural forces were also enlisted on the British side and the zamindar, Durgaprasad Roy Chowdhury of Bhowanipur, offered to engage Pandits at Kalighat from 20 to 24 October 'to propitiate the Almighty for restoration of peace and tranquillity to the country.'[66] Fakirs and Sanyasis, the other familiar presence in Indian political intrigue, are prominent in the records along with priests and Muslim preachers. Interestingly, ascetics often travelled in the train of the British army along with thousands of servants, families, and 'regimental women.' Even in the 1930s they are often a common source for the Criminal Intelligence Department.

Conclusion

India in 1857 was an information-rich world rather than a world starved of information and networks of circulation,[67] though this is the initial impression one gets from official sources. Sepoys and the British engaged in propaganda[68] and, for example, bilingual proclamations were published from a schoolmaster's press in Kanpur.[69] Allegedly, thousands of copies of the notorious Brigade Order of General Neil were distributed in English, Urdu, and Hindi upon its issue after 1857 and many other proclamations by the rebels and the British were circulating in the countryside even in 1887.[70]

The telegraph system proved no providential saviour during the uprisings of 1857. In terms of science and technology the Company's forces were not dramatically superior to the rebels except for their artillery and guns. The telegraph as an amazing feat of technology obscured the fact that it did not make the message it carried more credible; what was transmitted was potentially as inflammatory as by any other medium. Returns to indigenous networks of communication combined in 1857 with severe structural and spatial flaws in the original telegraph construction to make it less than useful to the Company's Government. While these flaws sometimes reflected shifts in technological perception and fashion, they testified to the continuing contradiction between technological myth and reality.

Knowledge or skill of telegraph technology was not transferred to the Indian sepoy and this was reflected in the many ways in which the system was cannibalised by the sepoys and rebels of 1857. That telegraph poles were used as makeshift cannons shows that culture did not circumscribe technological inventiveness, which was determined more by circumstance, knowledge, opportunity, and necessity. There might be a rare instance recorded when the sepoys tried to use the Delhi signal office but, upon failing, destroyed it. Luke narrates that on the afternoon of the 11 May 1857, very early on during the revolt, Ambala recorded movements of the needle as if Delhi was trying to get through but 'as no answer came to the usual question, ("What is your name?") they suspected that it was somebody unfamiliar with the apparatus.'[71] This brief glimpse illustrates the main factors inhibiting the rebel's use of telegraph technology. The sepoys were not trained to use it and lacked the knowledge to work it. The language used was English, which in most cases the sepoys did not speak. Finally, they had little reason to communicate with Bombay or Ambala: stations in British hands were useless for useful communication, while the stations in the affected areas were distances that were, perhaps, better served by human agency.

One of the reasons why the 1857 uprisings appear as the last stand of an older order or an example of primary resistance was the destructive attitude displayed by the rebels towards the telegraph. Fundamentally, the sepoys were the outsiders, and the bathos of the uprisings of 1857 is aptly illustrated by the incident of the mute instrument: technological knowledge and skill was never diffused and this was then used as an argument for western cultural superiority and technological acumen, and oriental backwardness and millenarianism. The telegraph system, designed to connect the cantonments of the interior to the colonial port cities, was useless to the agrarian unrest with which the sepoys had to negotiate. This illustrates the intrinsically colonial character of technology transfer under imperialism, one of the first and most comprehensive examples of its time: an extroverted infrastructure was very difficult to reconstitute inwards. The entire telegraph network in India was extroverted connecting the hinterland to the colonial port cities and had little, during this stage, to do with an integrated internal and national market. This facilitated Indian commerce on a global scale but questions of status, kinds of export, capital ownership and repatriation immediately became significant.

This chapter has concentrated on the sepoys to show how the mutiny occurred in a technological context and suggests that the character of the uprisings might have been different without the telegraph. The increasing deployment of Indian troops at the frontiers of empire precipitated the crisis. The cry of a 'Telinga Raj' or *'sab lal hoga'* combined with the anti-caste and the potentially radical nature of the soldiers' revolt. Trapped between agrarian and popular unrest, the middle and large landholders were forced to choose sides, and the vacillating roles played by some of the feudatories testify to this dilemma. The myth of the telegraph and more generally the protection afforded by industrial technology was mirrored by the invocation to Kali. Neither technology nor divine intervention saved many lives in 1857; the British struggled to victory without substantial telegraphic technological support while bullets killed the followers of Kali. The telegraph did not produce a more reliable message nor did it replace human networks of communication; it did not necessarily provide the quickest means to accessing information and its potential for containing information more effectively was not fulfilled. There was also no reason why rumour was any less rational in transmission; indeed it often won in terms of speed, or why the telegraph was any less irrational than human agency in its transmission of panic.

The telegraph did not save India, rather India saved the telegraph and the uprisings of 1857 saved the telegraph in India. 1857 saved the telegraph in India from an ignominious death, allowing reconstruction on a less experimental level. At a general level, it provided telegraph technology with the universal principle of duplication. The need to have alternate lines in case of interruption was unquestioned after 1857. The myth of the overland telegraph as saviour in 1857 renewed European and United States' interest and investment in telegraphy. Paradoxically, this interest was centred on submarine cable telegraphy as the least destructible of systems, given the unchallenged British naval dominance during the period. The events of 1857 were used to justify the need for an imperial telegraph system and the direct control by Britain over India, but the myth of the success of the landlines was used to justify concentration on submarine telegraphy as the most secure means of control. On a more human level, O'Shaughnessy's absence during the height of 1857 and the subsequent attacks on his style of management and technological acumen saw the 'father of the Indian telegraphs' quite discredited by the 1860s. Reconstruction after 1857 self-consciously and rigidly adhered to western models, instruments and expertise. In 1857, with the disruption to the telegraph, human connections and encounters gained vivid importance, and this account tries to capture this palimpsest going beyond monolithic narratives of technology and empire.

3
The Discipline of Technology

This chapter discusses the process of standardisation of telegraphic practice, and its consequences for business, styles of communication, and everyday life. It investigates issues of meaning and context, error and delay, language and time, to examine the definition of telegraphy after 1860 through social practice. It combines this with analyses of the recruitment and training of signallers, issues of departmental discipline, and the processes of professionalisation and institutionalisation. The process of standardisation of telegraph technology during the 1860s and 1870s contrasts with the period until the 1850s. Between 1830 and 1850 individual 'gentlemen of science' spearheaded technological innovation in telegraphy. The process of standardisation was in effect also a process of Europeanisation and metropolitanisation. After 1857 two trends emerged: first, scientific research and development on one hand, and industry on the other, were concentrated within institutional and industrial spaces in Britain.[1] Secondly, telegraph technology was continuously modified and tested by engineers and military men in India, yet no other recognition was given to these innovations except 'adaptation' to the particular climate and necessities of India, that is, reduced to innovations particular to India without global relevance. Institutional spaces concentrated resources and were legitimised as authorities through a variety of social and scientific strategies, including institutionalisation and professionalisation.[2] This chapter questions assumptions regarding the linear growth of telegraphy and shows how technology and society interacted to engineer a communication system. It also illustrates how governmentality had to increase in a period of laissez faire.

The main objective in the massive telegraphic reconstruction that followed 1857 was the rebuilding of the telegraph lines in such a manner as to make the system relatively immune to disruption in India.

This meant the development of a grid system within India whereby the failure of one line could be offset by alternate routes. It was noted that '...telegraphic correspondence all over India will be practically exempt from all interruptions, so many different routes will be opened up by which a despatch may be forwarded to a given place.'[3] India was also to be the crucial element in the telegraph web of the world. Expansion in this period followed not just the logic of commercialisation[4] but also the logic of strategic defence. More lines meant more alternative routes. The telegraph system would never again be completely linear or depend entirely on the rationally shortest single route.[5] Telegraph lines would be duplicated, triplicated and quadrupled. By 1859 the total length of the telegraph in India had rapidly risen to about 11,000 miles with 150 offices open to the public.[6]

The rationale behind the reconstruction of telegraphs lines

W. B. O'Shaughnessy's experiments belonged to the tradition of 'gentlemen of science' who were also seen as: entertainers and the workers of wonders.[7] There are two points to be made here. First, O'Shaughnessy's successor, Colonel D. G. Robinson, believed that innovations and attempts to set up local manufactories were wrong in principle. He argued that Europe had to be the model for Indian telegraphy and advocated instead the 'broad principle of striving to obtain those forms of telegraph materials which experience and the opinions of European Telegraph Engineers have declared to be the best adapted to the requirements of India rather than to experiment with the invention of our Indian *savans* [sic]'. He attributed the flawed working of the lines to a deviation from this principle; the department being 'inflicted with the Brooke bracket and insulator'.[8] Secondly, with the rise of the institutionalised laboratory, the actual implementers were no longer amateur scientists but trained engineers. The high degree of engineering training of military personnel and the militarisation of scientific personnel saw the implementation and maintenance of telegraphy transform into the preserve of the military, especially the Royal Engineers.

At both the level of invention and implementation, the inventing genius working with local skills and local knowledge was now anachronistic. O'Shaughnessy's immediate successor condemned the lack of insulation for the irregularity of transmission and also criticised the rigidly geometric principle behind the wires. Thus, the first telegraph system in India was stripped of its main claims to originality: non-insulation, and stark geographical rationalism in the construction of lines that ignored habitation. The telegraph in India was different from other systems elsewhere because it was constructed before the railways and cut across the countryside over the shortest distance. The telegraph and the railway, unlike in the US, were not the same lines and railway telegraphs and the state telegraph were competitors in India.

O'Shaughnessy refuted charges of flawed construction and accused 'dismissed employees and disappointed applicants' of ascribing the irregular working of the lines to 'wretched instruments and imperfect system.'[9] O'Shaughnessy was a difficult employer and an embarrassing employee: he blamed his employees for the technological crisis. He cited the fact that, during the rains of 1860, the line from Bombay to Indore returned a very low proportion of error, and claimed that this established the 'non-necessity for insulation.' He complained that his views were 'controverted by high authorities, and disputed, and opposed day by day by several of my own subordinates', and threatened that the 'man who assigns rain as a cause for interrupted work will probably have to look for employment in some other department.'[10] Contradictorily, he warned regular users to use the telegraph sparingly during monsoon.[11] He admitted that 'in the rainy season and the thunderstorm months, many interruptions occur from natural causes, which cannot be prevented.'[12] O'Shaughnessy denied the changing context of technology, its institutionalisation and industrialisation, and the emerging problems of labour management. Perhaps, his absence from India during 1857 contributed to this inability to understand the changed circumstances of the time. O'Shaughnessy left India on leave on 13 June 1860. He tried to interfere one last time in Indian telegraphy to protest against its reorganisation at the end of 1861, blaming 'matters for which the present administration of the Department is responsible':[13] the government replied, condemning the 'inherent defect of non-insulation'. Though a pioneer of submarine telegraphy, O'Shaughnessy was not consulted when the Parliamentary Committee drew up its final list of experts to be examined for the report on submarine telegraphy.[14]

There were several problems facing the Indian Telegraph Department after 1857. There was the question of the choice of routes: Lord Dalhousie's rigorously linear system had proved disastrous in 1857. It was now believed that a linear system was strategically harmful and financially unjustifiable. Rejecting a proposal to build a line from Gwalior to Jabbalpur through Sagar and Jhansi, the director general of the Telegraph wrote, '...the line would not pass through any towns of consequence.' The only purpose it would serve would be to open a more direct line of communication between the northwest and central and southern portions of India, a 'convenience that the present amount of traffic in the direction in question by no means justifies.' The line was not sanctioned. It was clearly noted that

> [i]n lines which had the recommendation of a joint political and commercial utility, it might be considered sufficient if the revenues derived from the offices equalled their working expenses, for the political convenience of the lines might be held as balancing the interest on the cost of construction and the expense of maintenance; but the case becomes altered when the proposal can have little beyond its commercial advantage to recommend it.[15]

The principle was that lines were to be built only if they afforded both commercial and strategic value. It is not that the department could ignore indigenous demand financially, but what it did was to use the cheapest and the most inferior materials. For example, the line from Agra to the indigenous banking centre of Mathura used Hamilton's 'half standard' conductors which had been withdrawn from all the major lines after having been found to be seriously defective.[16] It must, however, be noted that this was a temporary feature and the process of homogenisation and standardisation of technology brought parity on most lines.

Tariff and traffic in India

One of the primary means to control the volume of traffic and to subsidise government and press telegrams was the rate to be charged on private telegrams in India. How much this rate should be became a major area of debate in India. The Telegraph Department made its views clear on very low rates: the idea that cheap rates would attract general attention and lead to the incorporation of the telegram into the basic habits of the population, they argued, was not applicable to India. There was a threshold beyond which the telegraph was just not an affordable commodity to the vast majority of the people. In short, the Telegraph Department claimed that by lowering the rates, the government was merely subsidising the handful of those who could already afford to use telegraphs.

Commercially, the telegraph in India presented several problems illustrated by dramatic changes in the rates charged on traffic. The first phase in the 1850s with a charge on the numbers of words and a postal delivery charge was replaced by a low uniform rate in 1860. This meant multiple costs for an Indian. First, the cost of a telegram, calculated by linear, trigonometric, calculation of distance (these costs were calculated on a basis of slab or general rates, that is, for example, within 20–25 miles, 26 miles going into the next category of assessment). Second, the added cost of delivery by the post office further increased the cost. Telegraph offices were located close to the centres of European settlements and cantonments and delivery or receipt of messages outside this area was an extra cost for Indians communicating. These translations, writings, and retranslations cost extra money.

O'Shaughnessy's reported receipts increased almost 50 per cent, from Rs. 2,83,105 in 1858–9 to Rs. 4,23,991 in 1859–60.[17] However, press and government messages were subsidised by the telegraph system in exchange for guarantees given by colonial governments. These amounted to large annual sums. The total number of private messages increased from 101,164 between 1858 and 1859 to 1,70,566, which amounted to an increase of almost 70 per cent. The number of messages sent by Indians rose from 39,724 in 1858–9 to 71,554 in 1859–60. O'Shaughnessy was of the opinion that 'merchants and bankers, and the native community generally, can give as much work as the lines can ever perform.'[18]

The breakdown of private messages sent per telegraph circle in 1858–9 (Table 3.1) reveals the volume and pattern of inland traffic. It can be seen that by this time Bombay had overtaken the rest of India in terms of traffic and commercial importance. Bengal and North West province, with its headquarters at Agra, were still jointly administered from Calcutta after the 1857 disruptions, and totalled an amount that was just above that of the Madras circle. Next in importance was the Ceylon and South East Coast and the Punjab, Rohilkhand and Awadh

Table 3.1 Breakdown of private telegrams by administrative circle

Circle	Numbers (total = 170,566)
Bengal and North Western Province till Agra	39,315
Bombay	43,228
Madras	31,030
Central India	6,336
Indore	5,491
Ceylon and the South East Coast	10,766
Pegu	6,513
Punjab, Rohilkhand and Awadh	12,356
Scinde (Sindh)	5,660

Source: *Annual Report of the Telegraph Department 1859–60*, p. 9. NL

circles. Sindh, Central India and Pegu were the relatively recent areas of British penetration. Fewer lines and a smaller indigenous client populace slowed growth. The figures for Ceylon present a problem because Ceylon was independent and a part of the Indian Telegraph Department in phases. In 1858–9 it was a very new addition.

The problem continued to haunt the department: there were never enough regular users. After 1859–60, the government introduced a new flat rate for telegrams, perhaps, in the hope of increasing indigenous correspondence, and, therefore departmental revenue. It was also a move away from the earlier tariff system, which, as mentioned before, included very high hidden costs for the Indian correspondent. Comparing the relative merits of the first and second schemes, the Telegraph Department commented that the first scheme was 'as clumsy as the universal tariff is simple.'[19] The next scheme was to charge for distances actually travelled, supplemented by a booking fee, and delivered free of cost. After 1 October 1868 a uniform rate of Re. 1 for ten words was charged for the whole of India. In an effort to increase communications the rates were pegged significantly lower than rates in Europe, the US and Britain, and the existing rates in India.

The Telegraph Department recorded its reservations. In 1870 a further concession was made in regard to the address, in which three words were to be charged for the price of one. The director general recorded that he did not think the very low rate to have proved successful while agreeing, in principle, to a universal tariff for India. Though the volume of traffic increased, profits were offset by lower rates. By the 1870s the department was arguing that lower rates were not beneficial because the general growth rate over time was reflected in the traffic and the loss of money due to lower rates was only marginally compensated for, and that there was little substantial or dramatic increase in the traffic. Thus under the earlier tariff a ten per cent increase in number and value could be expected. This would have given the department Rs. 1,036,840 worth 310,582 messages but after the 1868 reductions they had 441,327 messages worth only Rs. 955,187. Even though the number of messages had increased by 56 per cent, revenues were nearly eight per cent below what would have been earned through the previous rate. The average yearly increase in revenue from 1860 to 1867 was nine per cent. This was claimed to represent the expected revenue increase due to 'improvements and extension of the system and natural normal expansion.'[20] Between 1867 and 1871, revenue increased by only about 11 per cent in three years, that is, between three and a half and four per cent per year. The director general concluded that the universal tariff of Re. 1 for ten words 'cannot be pronounced a success.'[21] Another significant fact about the pattern of indigenous usage emerges from the great distances telegrams covered in India. In Europe, telegrams often travelled between neighbouring towns. This was not true of India, where the bulk of the messages travelled between the greater trading centres, and between the producing districts of the interior and the seaboard.[22] It would appear that most of the messages sent were on commercial business.

The dilemma was simple: if the rates were low, potentially more people would correspond, but lower rates meant less revenue for the telegraphs. A sufficient increase in the volume of correspondents was never really achieved and increases in number were offset by lower profits. The Telegraph Department's views on very low rates were fiercely critical. The Telegraph Department argued that

> [i]n countries like Belgium, England or Switzerland a low tariff yields a large return, because the masses (unlike those in India) are rich enough to be able to resort to it, and because the average distance over which telegrams travel are comparatively small. The average distance for the whole of UK is probably less than 100 miles with the greatest distance a telegram can be sent in the UK is only 500 miles. The greatest distance a telegram can travel in India is 2200 miles and the average distance for all telegrams is no less than 1200 miles. ... It is obvious that the rates in India should be twelve those of England, other things being equal, but other things are not equal. In India ... unfortunately it is not a mere question of relative charges, but how to remove the large deficit which ought to be made good by those who use the telegraph, and not as at present by the ryots who are not likely to find any use in it in this country.[23]

In short, the Telegraph Department claimed that by lowering the rates the government was merely subsidising the handful of those who could already afford to use the telegraphs:

> None of the conditions exist in this country under which a very low tariff can be expected to succeed. The object of the reductions recently made in Europe is the extension of the use of the telegraph till it becomes, like the Post Office, a general necessity, to reach the masses in fact, and the large increase in the number of messages that follows more then compensates for the decrease in the rates of charge ... it is hopeless to think of reaching the masses [in India], and the only result of the introduction of a low tariff is to enable the very

limited section of the community that employs the telegraph to do so at less expense to themselves than formerly.[24]

In India while the department continued to expand, stretching from Burma to southern Persia, and subsidising trans-Indian traffic. This was a problem at the heart of the communications system. It was compounded when the average cost of a telegram in terms of labour and machine began to spiral downwards inexorably. Minimum profit guarantees and the telegraph monopolists tried to keep prices high as did the various governments. However, by the end of the century, the sheer downward pressure on prices and the consequent huge growth in information transmission made the state extremely vulnerable to and very aware of the need to contain, restrict and control the volume of information slowly spinning out of control. It is therefore unsurprising that information panics began to proliferate through the system after the 1890s.

Use and abuse of the telegraph: The Government Bank of Bombay or Cotton Bubble

That there would be a cotton crash in the 1860s was never in doubt; it was the timing of the crash that was a matter of debate. Goldsmid was told on his travels, while halting at a remote *sarai* in Persia, that it was generally acknowledged that the Civil War in the US was nearly over and that US cotton production would soon resume, to the detriment of producers in India, Turkey and Persia. The crash, when it happened, led to a general collapse of the share market climaxing in the crash of the Government Bank of Bombay.[25]

The telegraph played an important role in determining the timing of the crash and when news of it hit the market. Henry Collins was the first Reuters agent in India and arrived in Bombay in March 1866.[26] Reuters became the source of rapid information of prices in India, China and the Far East. There were around 118 public companies with an aggregate nominal capital of above £50 million that declared bankruptcy and rumours circulated rapidly of the impending crash of the large joint stock bank with the government as partner (see Table 3.2). Many of these were new companies in 1863, (see Table 3.3) and some of the commercial directors of the bank who were 'rigging' the market at the time were involved in their creation.

The most infamous being the Back Bay Reclamation Company, launched jointly by three firms, Ritchie, Stewart, and Company, Grey and Company, and the native Parsi merchant, Cowasjee Jehangir.

Table 3.2 List of new joint stock companies in Bombay, 1863

Banks	Capital (£)
Joint Stock Bank	500,000
Royal Bank of India	500,000
Bank of India	1,000,000
Broker's Banking Company	100,000
Total	2,100,000

Table 3.3 Table showing the run on the Bombay stock market in 1863–4

Year	Companies	Number	Capital(£)
1863	Banks and other companies	37	6,090,000
	Increased capital of old banks		1,225,000
1864	Banks and other companies	49	19,590,000
	Increased capital of old banks		500,000
1865	Banks and other companies	32	24,035,000
Total		118	51,440,000

Cowasjee was one of the directors of the bank, which had other Parsis on the board. The other two firms were also represented on the board of the bank.[27] The reports, pamphlets, and publications alleged that 'prodigious gains were made by those profiteering including some of the Government Directors.'[28] The bank then proceeded to give loans to firms that were already ruined, in order to try to recover their original loan, and individual directors were allowed to vote loans to themselves.[29] After the collapse it was learnt that the bank had been speculating heavily on the shares of the companies that had folded.

The bank had 16 directors at the time of the crash, of which six were government nominees. The Indian financiers on the board included Cowasjee Jehangir, Rustomjee Jamsetjee Jeejeebhoy, Limjee Maneckjee, and Premchand Roychand.

The shock of the financial crash and its scandalous nature was widely felt, and had important consequences. J. S. Mill wrote to Edwin Arnold at the *Telegraph*:

> It has always seemed to me that although the Bombay Government was only a shareholder in the Bank, yet as high officers of the Government were officially members of the Board of Directors, which did all the mischief, and as the Government itself not only neglected the duty of supervision, but, when repeatedly warned by

the Government at Calcutta, persisted in disregarding the warnings, and even withheld from the Calcutta Government the information it demanded, at a time when the disaster might still have been prevented from being complete; the Bombay Government is bound in morality and honour to indemnify partially, if not wholly, the shareholders who undoubtedly risked their money in reliance on the supervision exercised by the Government through the Official Directors. The case will shortly be brought before the House of Commons, and a word from the *Telegraph* on the subject would be of great importance.[30]

There are several points to be noted: first, that this was possibly one of the earliest instances in India of a crash based on telegraph information about international events and prices; second, sections of the bureaucracy were involved in the events, and the financial crash was carefully engineered over a period of three years of implausible buoyancy in the market, suspiciously ignoring common expectation of the impending crash. Finally, following Vicziany's work this analysis argues that the crash of 1864–6 allowed European firms to dominate the commerce of Bombay.[31] Though she rightly privileges the long-term factors changing the nature and extent of indigenous control and participation in the export trade, the short-term effect of the crash underlined the international and telegraphic nature of the export trade after 1864. The European firms were perhaps better placed to manage and process international prices through the English or European language-based telegraphs, and rigged the market to demonstrate this fact. Henry Collins, Reuters' representative in Bombay, regularly published Reuters quotations, which showed the price of different kinds of cotton from around the world exported to Britain. These quotations reported 'rapidly falling prices' for Indian cotton, and 'were widely regarded as a sort of death-knell for the Indian cotton merchants'[32] yet Indian merchants trusted Reuters and their equality in distribution of information to subscribers.[33]

Disciplining signallers and removing errors in practice

O'Shaughnessy confessed in 1856 to the 'anxiety of mind arising from introducing a new art and craft among many hundreds of people'.[34] Disciplining the signaller proved no easy task and the events of 1857 threw the entire telegraph establishment into disarray; many died, deserted, or were replaced with military recruits. The main amendments to the Telegraph Act XXXIV of 1854 were in Sections XI and XII: the first introduced heavy fines and imprisonment for the violation of secrecy and the disclosure of government despatches, and the second introduced fines and imprisonment for offering bribes to persons in the Telegraph Department. Delays in transmission and the loss of messages were frequent and inadequately regulated up to 1861.

Signallers quarrelled between themselves because checks on the time taken to transmit messages and fines for errors in messages made them competitors and enemies; even the most honest signallers disputed among themselves over where the blame for errors in messages lay. Other less rational examples were, for instance, a 'smart' or experienced signaller sending to a new or slow receiver who required many repetitions stopped sending, demanding a more experienced hand and insisting on the presence of the assistant in charge of the office. The receiving signaller frequently refused to do so, resulting in the 'stoppage of work for as much as a half hour at a time.'[35] Messages were lost during the night when the signaller, after writing the time of receipt and the name of the signaller, fell asleep only to be woken up by the repeated calls for acknowledgement from another station for a different message, which he acknowledged believing it to be the first message: 'the second message is thus lost.'[36]

Lines were 'switched over' without adequate notice, for example, when, after receiving a message from Benares, the Allahabad operator, without notice to Benares, 'joined over' or placed Benares in direct or 'through' communication with Kanpur; the Benares signaller sent more messages believing it to be Allahabad while Kanpur, receiving no notice of receipt of messages from Benares, did not record them. O'Shaughnessy reported that all correspondence between Calcutta and Bombay was halted by the head assistant at Sepree, a relatively unimportant node in the system. A signaller, a university graduate, 'being drunk, incapable of doing duty, and mischievously determined to prevent any of the signallers from performing theirs'. O'Shaughnessy complained of the difficulty in procuring the 'right quality of operators'.[37] Though this complaint appeared throughout the history of the telegraph, interestingly, because telegraph employment was still based on patronage O'Shaughnessy had a lot of applicants to choose from while later the same complaint disguised the relative paucity of applicants. It is important to note that potentially, any telegraph station could suddenly become the most important centre in the network if it ceased transmitting.

Corruption was rampant in the initial days of the telegraph. There were cases of delaying rival business telegrams, of changing the numbers actually quoted, there were systematic information leakages, intrigues within the department, and a huge circulation of telegraph messages between the operators themselves. For example, messages were sent by operators changing figures from five dollars per ton to four dollars per ton to two rival firms or changing the promised credit limit of a firm as promised by a London bank from two to 1000.[38] Sheeb Chunder Nundee protested against a temporary suspension for the 'extreme tardiness in submitting his accounts' that he had

> never been idle or neglected my duty, as FALSELY brought to your notice ... your kindness alone has risen me from a Writership of Rs. 20 a month to the rank of an Inspector on Rs. 200 [a month] ... I have sacrificed everything, and have more than once risked my life.

He alleged a conspiracy against him, which by falsely increasing the errors in his division, brought against him the charge of incompetence:

> The incessant interruptions which had taken place on the only section of my line between Sherghottee and Benares was not the result of any defect on the line, but that they were designedly caused by some of the hands to fulfil their dark designs against me, and in which they had very nearly succeeded had I not taken the timely precaution of placing a confidential hand to detect and point out which side the fault lay...

He recommended Ramcumul Chottorjee [sic] through whom he had been 'enabled to check the wilful interruptions during these two months'.[39]

A particular area of corruption was the opium speculations and this, according to O'Shaughnessy proceeded from the 'demoralization caused by the opium trade itself.' The instance of two dismissed signallers being employed by some opium traders to tap the telegraph line and set up an illegal office caused a sensation and had repercussions for the department as a whole. Their strategy was to get the information first and transmit slightly varied messages to rivals, and they tapped the line two miles from main road from Pune to Satara. As a result of this the department made it a policy to keep the lines close to roads and railways and visible public places.[40] Other cases involved bribes for information and for delaying the message of a rival firm.

O'Shaughnessy complained of 'scandalous neglect' by his own subordinates: the telegraphed 'No Demand' certificate from the accountant general, which he needed to board the ship taking him to Britain, reached him by post after 17 days, and after the departure of the ship on which he had booked his passage.[41] Given the increasing friction between him and his employees, perhaps, this was a case of wilful misconduct. He pointed out that

> anyone who has had to correct the first proof of an article set up by an East Indian or native compositor can understand ... how great is the liability to errors when the same agency is employed to write down from the dictation of an instrument, or to decipher its traces and dotting on a paper tape. The true wonder is that so few mistakes occur.[42]

He wrote of the 'lazy, apathetic persons we are compelled to employ in this country.' Dismissing claims that they were underpaid, O'Shaughnessy pointed out that in many cases

> young men who were useful and active while on low allowances, have turned idle 'gentlemen' when promoted, marrying at the age of eighteen, getting horses and buggies, indulging in low dissipation, and running deep into debt.

O'Shaughnessy was aware that he was unpopular with his employees. He complained he had been 'loudly accused' of 'habitual harshness and injustice' to his subordinates, and the press reflected this charge.[43] The government wrote to him pointing out that their 'approval ought to lead him to disregard the cavils and censorious remarks from the Press and private persons.'[44] He variously recommended the importing of US signallers along with Morse instruments, the recruitment of signallers from Britain, 'gradual conversion of the entire establishment into a strictly military organisation', more schools with more indigenous employees, and that the 'wisest course would be to sell our lines by auction to the highest bidder'.[45] Lieutenant Colonel Charles Douglas, his successor, recorded, on a telegraph establishment numbering 1193, excluding office servants and others on less than Rs. 10 a month, that 'misconduct, resignation to avoid dismissal, and dismissal, have been frequent.' Madras reported the lowest casualties, not because of any essential behavioural characteristics but because of the lower opportunity for employment and the low cost of living in that Presidency.[46]

A 'register of character': Constructing the ideal signaller

Disciplining these many instances required time but the blueprint was in place by 1861. A series of establishments were set up to monitor different aspects of the telegraph and its employees. The Complaint

Office examined complaints from the public, grouped around issues such as errors, non-delivery of messages (improper addresses) and the loss of messages, delay, and fraud. The Check Office of the Telegraph Department was introduced to audit service messages as prepaid telegrams and to reimburse them afterwards. A watchdog institution was established to oversee the minimum and maximum time taken over the despatch of messages, especially to Madras, Calcutta, and Bombay. A primary objective of this Fault Office during its initial years was the 'enquiry into breaches of office discipline', and it played an important part in upholding the 'discipline of the Department'. The Cash Checkers' Branch coordinated with the Complaint Office, which had the copies of all telegrams transmitted, to audit the payments of the telegraph offices to the Treasury against the actual number and words of messages.

Lieutenant Colonel C. Douglas complained about the 'absence of any register of character' in the department: 'no amount of good behaviour was remembered and, similarly, misconduct, however gross, was unlikely to be permanently remembered.' This increased a signallers chance of re-employment elsewhere or some time after dismissal. The solution lay in establishing a 'Character Book' in which 'all points throwing light, favourable or otherwise, on the character of the person, *however known*, would be recorded.' All reports by Deputy Superintendents, as well as carelessness or inattention traced through the Fault or Complaint Branches would be recorded against the facts 'favourable and unfavourable'.[47] A character roll for the telegraph establishment was drawn up with the following list of columns filled up:

(1) Knowledge of the language of the country;
(2) Character as regards temper with the natives;
(3) sobriety;
(4) any occasion for suspecting the persons honesty;
(5) special knowledge possessed, for example, ability to signal and receive by Morse or Needle;
(6) knowledge of electrical science;
(7) skill as a mechanic;
(8) natural ingenuity;
(9) amount and quality of education;
(10) health as affecting efficiency;
(11) peculiarities either temperamental or physical

A hypothetical form was filled out as an example: 'John Smith, East Indian, five feet four inches, dark hair and eyes, prominent teeth, sallow complexion, a scar on the right cheek, speaks slowly, and has a stooping gait'. Against 'Unfavourable Facts' were noted 'intemperate habits, no knowledge of electrical science, indifferent health', and the fact that the hypothetical signaller was 'very eccentric'.[48]

Disciplining the signaller involved policing his morals, movement, his self, and his space. The signaller was tested on several accounts in the early days of the technology ranging from basic office and signalling skills to a theoretical knowledge of electricity as well as mechanical skill in adjusting and repairing instruments. The Register of Character exercised a powerful influence over the conduct of signallers and it was noted that it was 'extensively known to the signallers how large an influence character, as recorded on their character sheets, would for the future have on their prospects in the Department.'[49]

Definition of the telegraph office: Disciplining of space

Signaller James Blacknight argued with the government about the 'impossibility of secrecy'.[50] If secrecy was insisted upon there had to be basic structures that allowed its maintenance. The transfer of telegraph offices from rented buildings to specially designed buildings reflected this necessity. There were several problems arising out of the enormous opium trade and speculations of the time, especially focussed on opium trade with China: first, the ease with which signallers off duty, often in an adjacent retiring room, could hear and therefore 'read' the messages and sell the information to interested parties and rivals. Second, there was no adequate control over the movement of signallers while on duty so that they had opportunities to divulge information to agents sitting outside, sometimes even on the pretext of going to the toilet. Finally, Receiving Clerks had to be disciplined into secrecy; they often divulged messages that were awaiting transmission. A minimum distance between the Receiving Clerk and the public had to be maintained. Doors, minimum distances, exits, and privies, all had to be designed and regulated.

Regulations laid down that none of the signallers off duty should be close enough to record the messages passing through office, and none of the signallers on duty should have any means to communicate with people outside of the signalling room. Lieutenant Colonel Douglas insisted, 'no one on duty in the Signal Room should be able to hear the messages received … the public must be kept at a distance from the Receiving Clerk to prevent the possibility of messages, after receipt, being read by others.' He instructed that the signal room should have only one access, through the sitting room of the assistant in charge, and this entrance must be easily visible. The signal room should have attached toilets accessible only from the signal room. Douglas stipulated that the relief of signallers, that is, the end of their shifts, should be 'as seldom as possible, viz, every twelve hours only, ... in the presence of the Assistant-in-Charge ... food must be brought and removed by a peon specially paid for this duty.' Signallers had to be, as a rule, moved away from offices when not on duty. So the practice of housing signallers in the telegraph office was discontinued. He directed all assistants-in-charge to write to him, with detailed plans, for the rearrangement of telegraph offices to ensure confidentiality. It was decided that the government should build its own telegraph offices as a matter of policy to ensure these conditions.[51]

Militarisation and Europeanisation of the telegraph: Employment and training of engineers

Most of the employees of the Indo-European Telegraph Department belonged to the military, as did most of the senior officers who succeeded O'Shaughnessy in the Indian Telegraph Department after 1860. The Indo-European Telegraph Department had no civilian directors up to 1880. Until l850, Indians and Eurasians filled the Uncovenanted services. Subsequently, the lower grades still remained their preserve but the service was 'augmented at the top by a large body of Englishmen, including several hundred Engineers, besides Forest Officers, the members of the Education, Telegraph, and other Departments, and so forth, who are no longer working, as the old Uncovenanted service, in subordination to the Covenanted Civilians',[52] reflecting the haphazard way in which the service developed.

Cooper's Hill Engineering College was formed to meet the demand for civil engineers in the Public Works Department. In the first half of the 1870s, civil engineers began a protest movement against inequality of pay and pension rules when compared with the Military Branch. The ensuing correspondence and newspaper reportage attracted the attention of the students of Cooper's Hill. It was noted, however, that 'in an agitation of this sort, the case is always stated on one side only, ... [yet] there is a real foundation for the depreciation of this service.' Though the government abolished the inequality of pay between the civil and military branches, the civil engineer had to serve 36 years before he was entitled to a pension. The government was urged to remove these inequalities.[53]

From the middle of 1860, a series of institutions were established to support engineers and act as important political lobbies for them. Institutions such as the Institutes of Telegraph Engineers, Electrical Engineers, Civil Engineers proliferated. However, by 1879 the Government of India was reducing the size of the large and powerful Public Works Department, which had important consequences for the future of Cooper's Hill. The worried principal wrote that he disagreed that the reduction in the Public Works Department's expenditure entailed a reduction of staff:

> The question at once arises for consideration, how far this will affect the conditions of the supply of young engineers for the service, and what modifications should in consequence be made in the arrangements now in force at Coopers Hill. But, first, whatever reduction of the Engineer Establishment may be determined upon, it may probably be taken for granted that such reduction will not be allowed to take the form of a stoppage even temporarily of the supply of young officers to the service.

He pointed out the many dangers to the service if Cooper's Hill was completely abolished and stated that its establishment was 'sound in principle' and that it provided a 'system of scientific education.' The complete stoppage of recruitment, he argued,

> would be very deleterious to the service as the promotion process would be halted. Any reductions should be offset by maintaining the basic promotion structure and the induction of juniors into the ranks. ... Even after the large reductions now contemplated, the operations of the PWD [Public Works Department] will still be of an extensive kind, and an efficient public service for their conduct will be as necessary as ever, although it may be smaller than the present service. But no service can be kept efficient in which production is suspended.[54]

The principal countered with the charge of haphazard recruitment in the past leading to the problems. He noted that the 'redundancy experienced is quite recent ... only two years ago the demand for engineers was still very high.' He complained that the pension rules were inadequate and officers and local governments colluded in keeping employed older officers 'so as to avoid making them face poverty, which their meagre pension promises after retirement.' He concluded that the 'present excess of engineers is due quite as much to this block at the top as to an excessive supply at the bottom.'[55]

The rapid and haphazard Europeanisation of the Uncovenanted Civil Service would have serious consequences for the future. The very large employment of signallers and other staff in Britain between 1866 and 1871 meant that between 1903 and 1908 there would be almost 40 retirements at a senior level and blocks in promotion plagued the telegraph establishment because they were all recruited around much the same time. There was a particularly large number recruited by the Government of India into the Indian Telegraph Department between 1864 and 1866. This would have disastrous repercussions for the service and promotions because most of these men would retire between 1903 and 1908.

Technological fashions: Hearing versus reading

The period 1860–70 witnessed the process of standardisation of various aspects of telegraphy. One of the frequently debated questions of the time was whether the eye or the ear was the more reliable organ for recording signals. Reading by ear, standard in the US, was a controversial practice in Europe. The Indian Telegraph Department originally used O'Shaughnessy's Needle instrument that used the eye.

The Morse instrument originally came with a tape that recorded signals. The director general discarded the tape and introduced the non-recording Sounder, an instrument by which the signal was received by ear. This was objected to before the Committee of the House of Commons on the grounds that the Sounder was undependable. In actual fact, British telegraphists preferred the recording instrument that used the eye in reading. However, the same experts abandoned the Needle instrument in 1871 in favour of the Sounder, which they had condemned in 1866 for being used in India. The older class of signallers who worked the Needle instrument popular in Europe and India dwindled rapidly, replaced by the Morse instrument and its signallers. The problems with the Needle instrument were: first, the lack of speed in signalling and the persistence of the vibration of the extremely sensitive needle; secondly, difficulties in reading the signals, and injuring the eyesight of the signaller; thirdly, they could not be used for automatic transmission; finally, the need to have two different alphabets after the introduction of the Morse system. It was stated that the 'sooner every needle instrument is displaced by instruments on the Morse system the better for the efficiency of the Department.'[56]

These shifts and turns in the mode of interpreting telegraph signals reveal the constant social tensions and pulls behind technological practice and the very contingent nature of both the machine and its use. The US operated a system based on the assumption of continuous electricity with an operator breaking circuits. The magnetic preoccupation of European science resulted in the Needle Instrument, which read by the eye and involved sending a charge to the circuit. Different paradigms inspired the two systems: the first was based on an imagination of interruptions to the field of electrical power, and, the second adhered to the notion of electrical flows via media.

The Morse code and other ciphers: Disciplining public practice

The previous sections dealt with various classes, sources, and strategies employed by the Telegraph Department in managing and recruiting its employees. However, the technology itself was in a process of evolution and change. Though the basic transmission template was Morse code, what was transmitted through it could itself be in code. Dictionaries and encyclopaedia of code were numerous since the early days of the semaphore and Claude Chappe. O'Shaughnessy originally designed his own code for the basic transmission template and claimed that it was more logical than those designed by Sir William Fothergill Cooke and Sir Charles Wheatstone. However, by 1850 he admitted the efficacy of the Morse code, which was being universally adopted. The fixing of the standard code of transmission did not mean that codification stopped. From the Agency Houses to the government, everyone brought to the Morse their own ciphers and codes. Thus, messages were doubly encoded: first, in the privately agreed upon cipher, and then in Morse.

A major issue between the public and the state was the use of these ciphers, 'generally composed of the names of heathen gods, ancient and modern cities, abstract qualities, etc.', which greatly multiplied the chance of error. The telegraph initially banned business codes, and messages composed of words of 'secret meaning' were strictly against international rules. They were later accepted 'on toleration' after 1868. Governments explained to the public that the code message became so obscure that it lost all context of meaning for the signaller. How was one to transmit correctly what one did not understand at all?

Perhaps, that was precisely what the public intended. However, signallers intensely disliked such obscurity because they were fined for errors in the transmission of messages. Though it was possible to trace an error down to the specific exchange between two signallers it was impossible to find out which of the two caused it. As a result both of them were fined. The department communicated to the government the 'difficulty and consequent distaste' that the signallers experienced for the transmission of cipher messages. The director general wrote:

> I refer to the class of cypher messages in which ordinary English words with concealed meanings are used ... [since] the fine for the error is shared between them [two signallers], it will be understood how sufficient a cause for wrangling and delay will arise during the transmission of such messages. Should the receiving signaller believe that an incorrect word has been sent to him – for example, 'dolt', 'spink', 'Crewe' – all three occurred in one message from a Calcutta firm, he insists that the sender is wrong while the sender insists that these are the words. The message delays others or is itself delayed.[57]

The government wrote to the various Provincial Chambers of Commerce in 1861 soliciting their 'general opinion as to the efficiency of the Electric Telegraph for commercial purposes.'[58] Their opinion was harsh: '... a merchant incurs a grave responsibility who acts upon any Telegram the purport of which does not tally with his preconceived notions, or, in other words, if a message reports an unexpected high or low rate of prices, it is probably a blunder of the Telegraph.'[59] Thus the general opinion seems to have been that the telegraph had to be tested against intelligibility of meaning and context. A common practice employed by European agencies was 'packing'. The fact that messages were charged on a minimum slab rate of 20 words allowed the sending of a message that was, in fact, many messages.[60] These agencies operated in west Asia and India until the end of the 1870s.[61] Reuters also employed this method in transmitting their bulletins.

During 1860–1, there were 130 complaints from Indian correspondents and only ten of those referred to mistakes with numbers. Indigenous merchants were cited as better correspondents than European firms because they did not hesitate to use devices and repetitions when it came to numbers, a very common source of errors. For example, they often repeated a number in terms of its multiple, thus, 'four thousand or twice two thousand', thereby eliminating all possible misreading. The bulk of errors occurred in the transmission of names and addresses and Indians, 'under the impression ... that greater attention was paid to the correct and proper transmission of messages addressed to Europeans ... several native senders were in the habit of adopting familiar English names such as Dr Green, Peter, Grant, John, etc., under which to send their messages'. The director general of the Telegraph Department confessed that 'it is undeniable that, under that practice, the errors in the addresses of their messages were materially reduced.' This 'practice of necessity' was prohibited when the rule demanding the real signature of the sender was enforced.[62]

The perfection of technological practice and the identification and removal of error concerned telegraphy after 1860. The telegraphs were constantly improved towards a perfectly working state, which was imagined ideologically and in technological idealisation but contradictorily constantly disrupted in reality. The fallibility of technology reveals the interface between the social and the technological and the fragility at the moment of transmission. Unlike the postal system, for the first time a message was not physically present but a translation bringing issues of authenticity, time, evidence and credibility into focus. The Indian case illustrates the complex experience of technological modernity.

Disciplining language and defining 'error' and 'delay'

Language was fundamentally transformed through the telegraph system. The initial charge levied upon each word in a telegram led to an almost inscrutable condensation of language. The reduction in tariff in 1860 saw an increase in revenue but the increase in revenue came less from the rise in the number of messages and more from the increased length of the messages as they were less cramped and more spelt out. This meant that the department also benefited from a decrease in the number of errors in messages, tabulated from the start of telegraphy.[63] Similarly, the allowance of three words in the address to be charged as one from 1 March 1870 aimed at reducing the number of undelivered telegrams by allowing more detailed addresses: in terms of revenue the strategy did not work.[64]

While specific government fiscal and departmental strategies often failed, the drive towards the evolution of a cryptic, even coded language is undeniable. Bennington argues, using Montesquieu that modern politics concerned with state power arose from the widening sphere of mass communication.[65] Perhaps, it was the nature of the post and telegraph systems to lend themselves as much to state centralisation as they did to subversion. The evolution of a cryptic, increasingly condensed language was as much a fiscal and societal expedient in order to pay less and communicate more, as it was a reflection of the changing metre and perception of time. The basic strategy of dropping articles, verbs, and other explanatory markers in everyday communication is an accepted element of language practice today. This was as much an understated trend from the late eighteenth century as it was an act of conscious self-fashioning. For example, the erudite and self-conscious exchange between two Cambridge scholars through the pages of the *Times* over the word 'telegram' versus 'telegrapheme'. The question was which was better Greek.[66] Less noticeably but more persistently, the 'telegramme' had quite wide currency in contest with the 'telegram': 'wire' was the easy and evocative way out.

The telegraph impacted significantly on language in a way that raised problems of context and meaning. Issues such as hearing, reading, secrecy, publicity, error, time, and interpretation enmeshed in the telegraph. Saving telegrams, repeating them, the elaborate records of error, institutions and offices established to scrutinise departmental functions, all concentrated on tracking down the source of error in a message. Yet the process of disciplining the signaller and training the public was only achieved over time. Mistakes in signalling were compounded by the cost of the telegram being fixed according to words, which saw brevity rise to absurd heights. The more compact and obscure the message, the less chance the signaller had of guessing what might have been intended. Telegraph guides emphasised 'it is of extreme importance to study the utmost brevity compatible with clearness', giving examples of how messages should be written and the acceptable degree of brevity without the loss of intelligibility.[67]

Signallers argued with the government about whose fault it was when a message contained errors and the government argued with the public about what constituted a correct telegram. The department castigated 'indiscreet condensation habitually practised' by the senders of messages. Such 'contracted' telegrams were unintelligible to none but persons between whom they passed. As a result, 'Office Assistants lose that control over correct transmission which the context of an ordinary message affords them.'[68] Telegraph guides suggested that 'the more intelligible [a message was] to the signaller the less chance of mistake.' They advised the use of words common to German, French, English or Italian; for example, words such as paper, maximum, sentence, statistics, moment, seraphic, communication.[69] Since the department gave compensation to the public for mistakes and errors in telegrams the disputes between the two were frequent and acrimonious.

The anxiety of authorial control, authenticity, and secrecy remained at the heart of telegraphic practice, although increasing automation led to the erasure of the human signaller. Yet up to at least 1920, signallers were the primary means of operating the system and enjoyed a privileged position. In the period 1855–70 they were the most important element, not only in transmitting the message but also in deciphering it. This interpretative role has been given little attention except within the terms of the institutional narrative of 'correction of error'. What constituted correct telegraphic practice was a matter of serious dispute between the public, the signaller, and the Telegraph Department.

The fallibility of technology

There are, therefore, two issues involved: first, mistakes arising out of codes that were added onto the basic Morse template, the cipher telegrams; secondly, mistakes that occurred in reading and transmitting the Morse code itself and the repetition of telegrams at different telegraph stations multiplied the potential for error. For example, Captain Matthews, telegraphing from Bombay to his agent Sheo Gopaul Dass in Indore, telegraphed, 'Sell all, and do not receive the horses. I am sending four thousand rupees', but, on arrival at its destination, the message had changed to 'Sell all, and do not reserve the horses, I am sending fourteen thousand rupees.'[70] Similarly, mistakes on messages transformed instructions like 'ship by sail' to 'ship by rail', involving a major change in transport costs and the bankruptcy of that firm, or 14,000 became 4000 or 40,000. Brasher gives a compelling example of misreading the dot-and-dash of the Morse code. The Times of 22 May 1867 [sic], published a telegram announcing an 'outbreak and suppression of mutiny at Great'. Re-writing it in Morse and changing the reading one gets Meerut; the word intended.[71]

There were problems with the technology itself. After his retirement, O'Shaughnessy wrote to the director general of the Indian Telegraphs defending the high percentage of errors in the transmission of messages by pointing to the 'prevalence of natural electric currents all over the India lines.' All it took was the transformation of a dot into an extra dot or even a dash or the shortening of a dash into a dot:

> Now in every thunderstorm, and even in storms hundreds of miles distant, the instruments are every now and then acting of themselves, and frequently and frequently for several minutes making dots and dashes and actual figures. *So that it is only by the context of words and letters that the work of the signaller can be distinguished from the effects of electrical disturbances in the air.* (italics mine)[72]

Similarly, Professor Ansted commented in a book published the previous year:

> The electrician places the wire as a means of communication, and at once receives a message from nature herself. ... An Aurora Borealis is seen–a magnetic storm is commencing. At the same instant the news is transmitted along the floor of the ocean by means of our wire, forwarded by no human hands, and in accordance to no human code of signals. Backwards and forwards, as if endowed with some strange vitality, the telegraphic needle is seen to vibrate, and the electrician must stand by powerless, trembling, like Frankenstein, at the monster he has called into life. The magnetic storm passes through the earth, and the use of the telegraph by man is for the time suspended.[73]

The Morse code was the fundamental divider of time. Dots, dashes, and spaces between letters and words were essentially a function in time. A recent study of the impact of the telegraph on United States' contract law, reiterates that 'perhaps misplaced dots and unwanted spaces were inherent in the technology.'[74]

The problems of interpretation and the tracking down of error led to several effects on telegraph technology and its practice. The areas of impact were: first, the quality and precision of the signalling and receiving instrument; second, the need to maintain a record of every transmission, ideally automatically, as a record independent of human agency; third, changes in the Morse code itself so that it became less liable to variation and error; finally, defining what legally constituted legitimate error. By 1870 several modifications in instrumentation ranging from Wheatstone's transmitter to Thomson's siphon recorder were undertaken. However, the operator remained the central figure and scontinued his magician's role of deciphering and interpreting electrical signals up to at least 1914.[75]

Writing about the recently invented siphon recorder, the *Daily Telegraph* enthused:

> Hitherto her [Electra's] messages through these long ocean cables have been taken with a mirror, so placed that its faint movements flashed changing specks of light, forming a language as difficult to learn as Assyrian arrow-heads, since the operator has need to know every tiny sign of earth currents, old signals oozing out, inductions, and many other interruptions, and to read the dancing dot clear of all these. The Siphon Recorder writing on ink is the way out ... a glass mouth and a tongue of ink.[76]

What is relevant here is the interpretative skill of the signaller – he has a hermeneutic power over the telegraph's language and transmission of information yet is powerless in overcoming the anonymity and erasure of his individual identity imposed by the system. He was the recorder of signs and the reader of portents yet he was also invisible, a mere means whereby electricity spoke. His very anonymity convinced the public of the confidentiality of the telegraph. Yet this was achieved over time and through a process of discipline. In triumphal celebrations of technology, operators are subtly erased as if there was no human agency in the system, or appear in a quaint chapter outlining early explorative exploits, heroism, and love matches between operators by means of the telegraph. Other histories focus exclusively on the operators and employees as a class in formation and use them as staple for labour history. This work attempts to overcome both preoccupations. O'Shaughnessy noted that the 'process of receiving a telegraphic message is one in its very nature liable to the misinterpretation of what is seen or heard.'[77]

Morse code changed as recording and transmitting instruments evolved. Operators decoded recordings on tape and wrote the message on telegraph forms, so the original Morse code was primarily interpretative. The widely prevalent practice of reading by ear further complicated the issue since sometimes no written record was kept. It was the Thomson siphon recorder that allowed a fully mechanical record to be maintained but what the symbols meant was still an act of interpretation. Wheatstone's punched tape was another breakthrough where dots ran vertically and the dashes diagonally with a perforated line in the middle. The retardation of signals was so high in submarine cables as to render the Morse code virtually unusable. So dashes instead of being distinguished from dots because of their length were now opposite in direction of the flow of current. Thus the dots appeared above the perforated line on the punching tape while the dashes came below this line. This was called the Cable code.[78] It was possibly the development of an automatic printer capable of translating the perforated Morse signals into ordinary type printed at speed on a plain tape, the telex, which removed the operator and was very important to high-speed wireless circuits in the 1920s.

Instructing the bureaucracy and the public in the correct degree of brevity to be used was a difficult task for the Telegraph Department. During the 1860s 'real' words were used to mean something other than their dictionary meaning. However, the dictionary code was supplanted by a code that used completely artificial words. Though various codes existed before the invention of the telegraph, by the 1920s it was stated that the 'compilation of codes involves more than the random selection of letters or syllables and the assignment to them of particular meanings. The successful compiler must be familiar with telegraphic practice; he must know wherein lie difficulty in transmission and the danger of mutilation.'[79] The double code, the basic Morse code and the cipher code, and their proximate yet opposite positions led to more than 3000 codes compiled. Originally, the rule of pronounceability was adopted towards codes but the number of easily pronounceable words being exhausted, 'barbaric combinations were presented.' In 1920 an international conference met to discuss, among other things, the problems of transmitting words like 'NHAZBCLYOZ' and 'BYMGAVRASMB'. Competition between operating companies broke down any possibility of strict enforcement of the 'pronounceability rule', which was very hard to apply in the first place in view of the possibilities of eight different languages (German, English, French, Spanish, Italian, Dutch, Portuguese, and Latin), the lack of linguistic ability or knowledge of telegraph clerks, and particularly the lack of a precise definition of the term 'pronounceability'.[80]

Conclusion

This chapter argues against the seamless growth of telegraphy: telegraphy was fashioned through experience that was often contradictory, but that carried a strong authoritarian motif. The telegraph was visualised as a means of exerting power; it exercised exclusion and disciplined social practice within its establishment. The experience of telegraphy introduced anxieties about meaning and credibility and the fallibility of technology multiplied errors and increased its vulnerability to panic. This disciplining was universal through the telegraph establishment and the training of recruits was repeatedly reorganised along with recruitment policy. Simultaneously, the metropolitanisation and Europeanisation of Indian telegraphy between 1867 and 1871 created internal management problems by the turn of the century. The idea of the perfect telegram, transmitted via the perfect line to reach the ideal addressee, shaped the elaborate architecture of corrections and the design of telegraphy. This case study of management strategies in the Indian Telegraph Department is also a discussion of the imagination of telegraphy. This disciplining of language, labour, office space, technological and signalling practice, standardisation and improvement of technology, demonstrated how governmentality increased even during the period of laissez faire liberalism. The very nature of the technology required a greater degree of state control and centralisation and the need to deploy a much greater bureaucratic apparatus than would be warranted in an age of laissez faire. Telegraphy became a central part of the state yet this chapter also shows how people's practice, the rate charged on telegrams, signalling practice, and the anxieties of the state interacted in the ongoing process of definition and disciplining of telegraphy. Language changed, as did the character of the ideal signaller, and issues of control and trust diffused throughout the system. At the same time, strategies of subversion multiplied and went global under telegraphy.

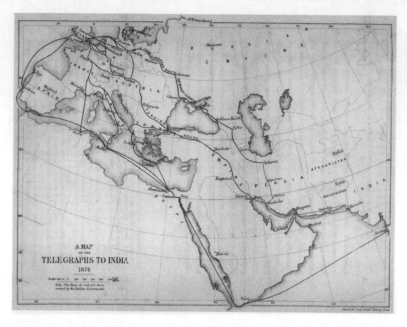

Map 4.1 The overland routes to India
Source: Personal collection

4
Making the Twain Meet: The New Imperialism of Telegraphy

The theory of 'rupture' and the stages of metropolitan capitalism divided British rule over large parts of the world into different periods. Versions of the expansion of Europe since the sixteenth century show that the international system grew in internal and central strength through the exploitation of, and at, the periphery.[1] Economistic explanations of the thrust for imperial expansion in the nineteenth century were criticised as mechanical and Eurocentric.[2] Gallagher and Robinson allowed the dynamics of a locality, region or country to return to what had been predominantly a story of metropolitan industry and capital.[3] British expansion after 1870 occurred in the context of industrial decline in Britain:[4] did the Empire, carved out of the 'bargain basements'[5] of Asia and Africa, exhaust Britain? A strand of historiography sees this as a period of British industrial decline and therefore of imperial defensiveness.[6] Expansion after 1840 was not the product of an expansionist British policy but of the need for 'military security, for administrative efficiency, or for the protection of indigenous populations on the frontiers of existing colonies'; 'to this extent, late nineteenth century imperialism was merely the continuation of a process which had begun centuries earlier.'[7]

Studies of the relation between the New Imperialism after the 1870s and science and technology seem few and far between in contrast to more political and cultural studies. The excentric and non-economistic strand of historiography went in the opposite direction to argue that technology had little role to play in the imperial expansion after 1870. The word technology appears just twice in D. K. Fieldhouse's history of the economics of the British Empire,[8] while others mention it only in passing.[9] Firearms, compasses and sails, were the technological edge that when combined with the rigorous organisation of its troops made the first empire, with its naval superiority, invincible.[10] Steam, electricity, and petroleum were merely changes in degree of empire rather than constituting a fundamentally different empire. After Cipolla and Marshall's[11] engagement over the issue of culture and military technology, Headrick analysed the establishment and evolution of the steamship and the gun. In doing so he did not stray outside the narrow confines of the debate. Other debates that have emerged concentrate on the collection and classification of flora and fauna, as well as on the collection of data at the margins of empire. Set against the idyllic backdrop of a tropical paradise, the explorer-botanist networked with the metropolitan expert to conduct distinguished exercises in the production of knowledge. Such studies confined themselves to institutional spaces, correspondence, and metropolitan practices.[12] Headrick distinguished between the means and motives that constitute causality; motive being the word used by Robinson to defend the eccentric thesis against the charge that this view claimed that the mentality of the man-on-the-spot caused imperial expansion, and stressed the need to study the hardware or the means of expansion.[13] Headrick demonstrated how crucial changes in the design of guns and ammunition allowed small units to overcome numerically superior but hopelessly under-equipped armies; technologically combative indigenous resistance declined in the case of Africa, as perhaps, in any similar situation, the further the interior was penetrated.[14]

Churchill noted that the significant factor was the 'arms of science', which, for example, perpetrated bathetic encounters between the assegai and the Maxim gun. Battles became automatic encounters where the 'mere physical act' of loading the Maxim became 'tedious' while 'all the time out on the plain on the other side bullets were shearing through flesh, smashing and splintering bone ... valiant men ... were suffering, despairing, dying.'[15] The casualties were absurdly lopsided; for example, 11,000 Dervishes and 48 British soldiers were killed at Lord Kitchener's victory at Omdurman. Technology generated within the laboratory and the metropolitan factory found the empire a site for implementing and cultivating these experiments. The 'dum dum', a soft-nosed bullet designed to mushroom on impact leaving fatal exit wounds, was not used in European conflicts but reserved for use largely in Africa and Asia.[16] However, one is no closer to understanding the causes behind imperial expansion in the late nineteenth century if one does not ask why these fights were occurring. Locating agency at the periphery, Robinson initially argued that the 'Boxer rebellion of 1900 provoked the Russians to occupy Manchuria in much the same way that

Arabi's rebellion led to the British occupation of Egypt.'[17] He ignores the landing of the telegraph cable in China or the railway scheme through Manchuria to Vladivostok. Robinson, more recently, returned to the hardware of empire and admits that the

> locomotive, along with the steamship and telegraph, shrank time and space within and between regions and seemed to bring almost everywhere within striking distance of Europe and lend to it new economic and strategic significance ... the railroad was not only the servant but also the principal generator of informal empire; in this sense imperialism was a function of the railroad[18]

A particular communication system, the telegraph, predominantly modified notions of time and distance and therefore constituted the hegemonic determinant of the relations of production and power. The imperial web of overseas finance that emerged after 1870 was mirrored by the expansion of the overland and submarine telegraph system: the telegraph system provided both the means and a primary locus for the financial and imperial expansion of Britain after 1850. The telegraph formed the basis of a heightened order of imperialism in India after 1857. Obviously, to Victorians, it must have appeared an astounding age, dazzling yet comforting in the mechanical solidity of the railway, steamship, and telegraph. In the propaganda for an imperial mission, science and technology played an under-publicised but vital role. Imperialism in the late nineteenth century was nothing new but fundamentally different from all previous forms. The exocentric or continuous picture needs important qualifications. First, the notion of British industrial decline and Britain's increasing domination of southern or City finance should be reconsidered in the context of the diversification and specialisation of industry and industrial expertise, and the proliferation of turnkey projects, especially since British industry dominated the export of technology until the 1880s. Similarly, northern capital and scientific training dominated much of land and submarine telegraphy until at least the 1880s. For example, Sir John Pender, aged 57, was chairman of the newly consolidated Eastern Telegraph Company in 1872. Born in 1815 in Scotland, Pender went to school in Glasgow and established himself as a successful cotton trader in Manchester. Pender invested in cable technology and became the head of the cable empire. Known as the 'Cable King', he influenced imperial finance and British politics during the period 1860–1900. This highlights the ability of industrial and northern finance to penetrate the City and take the initiative in the manufacture of imperial policy; and Pender often played very dirty.

This monograph revives the notion of a New Imperialism in the context of scientific and technological change after 1850. In 1866, Gladstone asked Lord Derby 'What is the use of these colonial dead weights which we do not govern?' Though the reference was more specifically to Australia and Canada, the sense of frustration with the entire empire was clear. However, Arthur Knatchbull-Hugessen, the Liberal spokesperson for the Colonial Office in the House of Commons, clarified in 1871 that the Liberal government had no desire to dismember the Empire but that the 'Government wish to retain them [the colonies etc.] bound to this country by ties of kindred and affection.'[19] These ties were strengthened, transformed and physically implemented by the technological triad, and in particular by the telegraph. The various ways in which the technology served, sponsored and promoted the British Empire will be examined in this section: the telegraph itself as a source of information and propaganda; issues of censorship and control over the press; telegraph personnel as sources of intelligence; how it served as a symbol of British technological superiority; how the technology radically transformed the map of which places were important and which were not; how it led to a fresh race for territorial expansion and penetration; how it served to penetrate, exert central control and extend political sovereignty. Finally, the telegraph provided a field and context within which newer strategic imperatives such as control over petroleum emerged towards the close of the nineteenth century.

The Great Game: The telegraph to India as imperial defence and propaganda

Telegraphic connection between Britain and India was an imperial concern before 1857. Sir Robert Peel, Prime Minister of Britain, received a proposal to link Britain and India by land and sea telegraph in 1845, and telegraphy proved very effective in serving the cause of British diplomacy. The Sublime Porte invited Lionel Gisborne, who had submitted the proposal to link Constantinople and Alexandria as a preliminary step towards a telegraph to India, to meet with him in Constantinople in 1854. Negotiations were conducted through the British government, and, specifically, Lord Clarendon. Istanbul was connected to Vienna and all European capitals by 1855.[20] From these early negotiations, the European and Indian Junction Telegraph Company was born. It subsequently became the Red Sea and India Telegraph Company on

29 August 1857. That the negotiations with Turkey started so early is unsurprising in the context of the Crimean war, which served to underline telegraphic communication as a requirement of state with an overland line from Varny to Balaclava. For the Ottoman Empire, the telegraph meant an opportunity for ostensible modernisation and autocratic centralisation that was eagerly grasped. The Ottoman government allowed British expertise and British signallers to build and operate the line from Basra to Belgrade. It was formally agreed that in exchange for the reservation of a line primarily for Indian messages to and from Britain, the Indian government would provide part of the finance. Turkish eagerness to modernise, combined with Britain's interest in creating a diplomatic situation, was to prove useful in not only the construction of the telegraph but also in securing Turkey to the British sphere of influence, especially between 1850 and 1870. Abdul Hamid closed the constitutional parliament in 1878 and used the telegraph to centralise the Ottoman Empire.

It was important, especially after 1857, that Britain be telegraphically joined to India, 'the greatest of her dependencies'. This was an 'imperative', an imperial 'necessity'[21] for an empire that was, after 1857, to be ruled from London and for the British mercantile community whose enormous business was dependent on their eastern possessions. The adage of all roads leading to Rome was to be illustrated through an international network, which would have as its central nodes, Britain, specifically London, and India. By the 1860s strong arguments were put forward by George Balfour in the *Times* for linking Assam in India, Burma, and Schezuan in China. The demand for labour on Assam tea plantations, an increasingly high investment industry, could be supplied by supposedly cheaper and more malleable Chinese labour while sparsely populated parts of Burma and China could be settled by Indians. Tea from Assam would be bought in west China that now had to get tea from the eastern provinces. Furthermore, Rangoon would be a better seaport for exports and would prove crucial in bypassing the increasingly expensive and turbulent metropolis of Calcutta with its growing nationalism and Bengali lawyers tentatively starting litigation in favour of labour. Balfour argued, 'Assam is within 150 miles of the most populous province of China. ... Indeed, if we can break through the narrow barrier which separates the Province [of Assam] from Sze-Chuen, it would become one of the most important of all our Asiatic possessions, by reasons of powerful mass of labour, which that China Province could and would pour into Assam.' He perceived free trade and the abolition of customs tariffs as the primary means of penetrating Arabia, Persia and the coast of Africa. Free trade was also to be the means of securing China to the British sphere of influence. This ideology was to be the 'binding liberalism between Oxus and Indus' which Balfour believed to be the best mode that could be adopted by Britain for 'counteracting the evil designs of Russia in surrounding China, occupying Turkestan, and occupying Kashgar.'[22] Commodities were not just factors of trade but were essential elements of imperial propaganda. Balfour pointed out, '... once let the bazaars of Cabool, Kandahar, and Khelat be filled with Russian goods, and these goods probably penetrating to the hostile tribes on our Punjab frontier, then will commotions and pulsations be felt throughout all India ... the presence of a bale of Russian goods on the borders of India will prove to be powerful illustration of the extended and over extending influence of Russia.'[23] The immediate context for anxiety was the Russian entry into Kokand and Yarkand, which permitted Russia to command the ancient overland route connecting China with Turkestan.

After 1857, a concerted movement began to link Britain and India by more than one line. The first International Telegraph Conference in Brussels on 30 June 1858 began the process of securing European co-operation and the homogenisation of rates. Almost simultaneously, Britain was to be linked across the Atlantic with the US, which was rapidly emerging as an important consumer and producer. Growing French and American involvement in China and Indo-China could be serviced through the extension of lines through India: Britain would provide this service. Japan was expanding its sphere of influence in the east and emerging as an imperial power. To protect their investments in China and Indo-China, Britain, France and America needed rapid access to these regions and information about prices of goods within their spheres of influence. Furthermore, the threat of Russia, whether real or imagined, prompted an encircling movement where the Mediterranean, Turkey, Persia and Afghanistan connecting with India resisted Russian expansion to the south. India was a vast military reserve that was crucial to the empire and to resisting any attempt by Russia to enter these regions. British control over India would ensure continuation of communication between the west and east. The Great Game being played was not amusing when the huge expenditure of lives and money is considered. Kim's Lama and the Russian agent reappeared throughout the history of western engagement with the east and continue into the present.[24] The spectre of Russia impelled the need for a telegraph network that would include but contain Russia and secure India to Britain. It was not just a question of profits and economics in this period but also one of populations and men. These huge imperial militaristic machines needed enormous reserves of human life. India had a population that grew from 255.2 million in 1867–72 to 305.7 million by 1921. Three quarters of the total inhabitants of the British Empire including Britain lived in India.[25] The control over populations served not only as propaganda, threat and potential military reserve, but also in practice provided labour. Britain managed the demand and supply of labour, which was enmeshed with industrialisation, capital flows and daily information about prices of goods. Indian labour was sent to Africa, America and the Far East. The telegraph network coordinated this global movement of money and people.

Telegraphy, more than anything else, ensured the symbolic presence of British imperialism. The telegraph was the supreme celebration of the scientific empire that Britain promised. It served as the physical reminder of the apogee of rationalism, technology and science that the British Empire claimed to represent. The celebration of science and empire was fraught with difficulties and the ruled, those necessarily categorised as irrational and unscientific, watched and waited. The success of the telegraph project was not a matter of course and the subject peoples watched its development without neutrality. Failure of the project would have meant a crisis for imperial propaganda and prestige. It is important to remember that the telegraph project was a symbol of the empire, and during its early period, a spectacle watched with curiosity as well as scepticism. It was noted that the 'majority of the natives did not believe in the working of the telegraph lines at all and maintained that we "were cheating them out of their money without being able to send a single message!"'[26] Telegraph technology was political in every sense of the term; it was both the meaning and content of the British Empire.

Connecting Britain and India by water: The Red Sea bubble

So essential seemed the task of connecting Britain and India after 1857 that the government entered into a contract it was soon to regret. The hastily set up Red Sea and India Telegraph Company was guaranteed by the government on a capital of £800,000 to build a line connecting Suez, Aden and Karachi. R. S. Newall and Company, were the contractors for the construction and installation of the line. Gisborne and Forde were the engineers, and Siemens and Halske the electricians. Lionel Gisborne secured a monopoly from Egypt to work the landline in Egypt between Alexandria, Cairo and Suez. The Secretary of State wrote to the government of Bombay and the British Agent and Consul General in Egypt, Brigadier Coghlan, that 'when we used the term "suspension" we did not mean complete suspension' of official involvement but that they referred to Coghlan's recommendation to appoint an officer chosen by the government of Bombay 'to enter into negotiations with various tribes or chiefs in whose countries the proposed stations are situated, and to arrange for the transfer of ground on lease to the Company'. The Secretary clarified that the British government was

> [u]nwilling that any negotiations should be entered into in their name, and most specially, negotiations for the cession of territory to the Crown but they are willing that you should facilitate the operations of the Red Sea Telegraph Company in any manner that is not likely to involve the Government in disputes with native chiefs or otherwise create political embarrassment.[27]

The informal expansion of the British Empire for telegraphic purposes soon saw it formally involved in almost all parts of the globe. India, as usual, provided labour and expertise: R. L. Brunton was appointed General Superintendent of the Red Sea Company[28] and Captain Playfair, also of the Indian Telegraph Department, was recruited for the 'establishment of telegraphic stations on the coast of Arabia.'[29]

The immediate problem facing the constructors was surveying the depths of the Red Sea. The original plan to connect Ras-el-had and Karachi had been abandoned because of the great depth of the Red Sea, and instead turned to Muscat.[30] Finally, complaining about the impossibility of laying a cable until the line had been accurately sounded, the Company chose to connect Basra and Karachi.[31] The Government of Bombay prematurely requested permission from the India Office to allow telegraph offices to receive messages for Britain.[32] The Company announced the opening of the line between Alexandria and Aden[33] and by June of that year they claimed to have 'succeeded in laying a submarine cable down the Red Sea, and that telegraphic communication between Alexandria and India is complete.'[34]

By November 1859, the project had collapsed and the Company wrote to the British government announcing a series of failures both east to India[35] and west to Europe.[36] Imperfectly surveyed, the cable too slightly built and being laid too taut, the project proved a colossal failure. Furthermore, the warm waters of the Persian Gulf led to the rapid degeneration of the gutta percha covered cable, which was attacked by microbes and boring worms and the project had to be abandoned by

1860.³⁷ The failure of the Red Sea and India Cable Company with its huge financial investment raised mainly from British small investors in the form of shares was a serious blow to the imperial telegraph project. The government contracted to pay interest on the outlay and this was continued as a charge of around £36,000 a year until 1908. However, all was not lost and John Pender bought the Egyptian telegraph concession from the Red Sea Company a few years later. For the time being, an alternate and secure route to India had to be found. The same year a book was published anonymously criticising the encroachment of monopoly. It claimed on behalf of the submarine cable:

> Young as I am, however, few have undergone more suffering, and been subjected to greater cruelties during a long life, than I have in my short career. My severed and scattered limbs, now lying at the bottom of the ocean, in various parts of the globe, bear ample testimony to my ill usage.³⁸

Supporting Gisborne, it suggested that Indian rubber was a better insulator than gutta percha and criticised the monopolisation of the cable manufacturing process

> … how injuriously the Gutta Percha Company's virtual monopoly has acted upon the interests of all submarine telegraphs. To this company alone belongs the right of making and selling gutta percha in any way and in any form whatsoever. They raise and lower the price of the material at will, and from their charges there is no appeal. If any enterprising member of the House of Commons wishes for the strongest illustration to be found in this country of the evils which result to enterprise from these patent monopolies, he has only to enquire into the privileges enjoyed by the Gutta Percha Company, and then see how these have acted upon the cause of telegraphy throughout the kingdom.³⁹

Gutta percha was introduced as the main insulating medium in 1850 and rapidly became the staple of submarine telegraphy, although many advocated the use of Indian rubber.⁴⁰

The failure of the Red Sea and Atlantic cables led investors and engineers to doubt their viability. Between 1844 and 1855, the telegraphs built were primarily landlines, and, as in America, often followed the railway. In India the railways and telegraphs were divergent projects. This might have been due in part to the relatively late introduction of these technologies in India under different forms of control. The government built the telegraphs while the private companies built the railway and telegraph system. O'Shaughnessy proposed underground and underwater telegraph lines but was instructed by the government to construct landlines. Despite this, his enthusiasm for underground and submarine telegraphy continued unabated, so much so that he joined Morse on one of the main cable-laying expeditions in the Atlantic. Perhaps, having explored most of the world, what fascinated and challenged the Victorian men of science was what lay below the surface. Oceanographic exploration proceeded hand in hand with submarine telegraphy. Authors like Jules Verne and Arthur Conan Doyle articulated and popularised these ambitions at the level of fantasy and exploration fiction. At the same time attempts were made to join Britain and the US. However, the cables snapped throughout the early part of this period leading to the spectacular failure of 1858. The amount of money the telegraph companies could raise fluctuated with the fortunes of their lines. After the initial failure of the submarine cables, prices of shares for landlines spiralled upwards.[41] Sir John Pender had to invest his personal fortune to continue the attempt to connect Britain and the US as no money could be raised after the initial failure of the Red Sea and Atlantic submarine cables.[42] Cyrus W. Field had problems on the other side of the Atlantic. British expertise and the greater availability of risk capital in Britain saw the initiative pass from Field's hands into theirs.[43]

Connecting Britain and India by land: Issues of politics and sovereignty

The Indian uprisings of 1857 and the failure of the Red Sea Cable in 1859 re-emphasised telegraphic connection between India and Britain as the imperial imperative. Engineer Josiah Latimer Clark, investigating the damage done to the Red Sea cable between Suez, Aden, and Karachi, reported in 1862 that the cable was impossible to repair. The military engineers in the Indian Telegraph Department surveyed overland routes to Europe. Work on a landline through Turkey to Belgrade was started to connect Britain and India. In the 1860s negotiations were started for landlines connecting India with Europe. With the expansion of the telegraph system came the second phase of imperial penetration. Persia, Afghanistan, Turkey and the Mediterranean proved crucial to the maintenance of the lines, and from the 1860s the telegraph engineers spearheaded the diplomatic negotiations. Indeed, the second phase of imperial domination and expansion occurred under the flag of technology, especially the telegraphs. The Turkish government constructed a line between Baghdad and Constantinople, via Scutari, Angora, Diyarbakir, and Mosul, and an agreement was reached between the British Consul and the Sublime Porte for special wires to be reserved for traffic between India and Europe, as well as for the extension of the telegraph line from Shatt-el-Arab at the head of the Persian Gulf. The Anglo-Turkish Convention of 1864 proved invaluable in securing Turkey within the imperial communication network and for ensuring that there were special lines reserved for traffic from India to Britain. Dawud Effendi, Director General of Telegraphs in Turkey, was extremely efficient in constructing and maintaining the lines. The strategy adopted was to employ Arab horsemen and personnel for protecting and building the lines. This provided local support for telegraph lines west from Baghdad.[44] The line passed from Constantinople through Belgrade, Munich and Paris, to London. By the 1870s, another landline passing through Italy and France was built. Another line following the Persian coastline had to be constructed because the regions east of Baghdad proved far too inimical to British penetration. Siemens' Company contracted with the British government to build a third landline and the Indo-European Telegraph Company was launched. The line was completed in the 1870s and went through the southern part of the Russian Empire from Odessa to Jitomir, Warsaw and Berlin, to Britain.

Mid-nineteenth century coastal Persia was yet to belong to any clearly centralised state, and similar to the pattern of eighteenth century India, governors in outlying provinces were powers in their own right; satrapies, in short. Contributing to the difficulty for the British were the different tribes such as the Kurds, with their fierce sense of independence, and a border not clearly defined. In 1863–4, Lieutenant Colonel F. J. Goldsmid (future director general of the Indo-European Telegraph Department) along with the Colonel Patrick and Captain Charles Stewart (the former was director of the Indian Telegraph Department, and was the officiating head during 1857), Sir Charles Bright (a telegraph expert of repute, subsequently involved with a number of cable companies), Latimer Clark (another electrical engineer soon to be involved in the Atlantic cable project and subsequently knighted), H. I. Walton (soon to be the head of the Persian Gulf Telegraph section), and several Royal Engineers and Lieutenant Stiffe of the Royal Marines, surveyed and laid land and submarine lines on the Persian and Baluchi coast. Goldsmid was appointed the Political Agent accompanying the telegraph cable team for the laying of the Persian Gulf line. His task was to arrange the Persian Convention that would deal with issues of sovereignty as the telegraph relentlessly pushed on towards Europe. Another reason behind the Convention was to persuade the Ottoman government to officially agree to employ clerks who had a working knowledge of English.[45]

In the draft of his report to the Bombay government on his journey from Baghdad to Constantinople he commented on the following subjects: (a) the route, specially with reference to Telegraph purposes; (b) the natural products, and means of obtaining water and supplies; (c) the inhabitants: with some assessment of the larger towns, and matters of political and general interest.[46] This ordering of priorities shows the political importance of the telegraph in the hierarchy of imperial interest. More revealingly, he complained that he had been authorised by the government 'to render every assistance in my power towards the establishment of the great line of the Electric Telegraph. The consequence has been that the latter work has absorbed the former',[47] that is, the Mekran enquiry into political penetration in southern Persia by Russia, which was a political task. He would later claim in his book that 'it will also have become patent that the difficulties encountered [in laying the telegraph] were rather of a political than a physical character.'[48]

Goldsmid's complaint highlights the fact that the telegraph had crucial political dimensions and was to form the basis of an intricate and inherently fragile world order based on issues of sovereignty, territory, apportioning of rates and tariffs, who would build and work the telegraph, and how it could be rendered safe from interruption.[49] Ironically, the director general of the Indian Telegraphs complained to the government a few years later that 'both General Goldsmid and Major Champain have been incessantly engaged on diplomatic negotiations which have kept them busy and too far removed for them to attend to the working of their lines.'[50] Goldsmid noted:

> Political difficulties presented themselves on the Arabian coast ... A question had arisen as to sovereign control over the isthmus and inlets now under utilisation of the telegraph scheme. However peacefully traversed by the wires on a chart of Eastern seas, the reduction to actual practice of a Western hypothesis was quite another affair: and when the real scene of action was approached, the Arab fishermen and inhabitants concerned were reluctant to bestow their friendly offices on comparative strangers without, at least, the guarantee of some substantial return for a manifest privilege.[51]

Reporting on the threat of an uprising in Mesopotamia, Colonel Kemball noted that 'security has ... been re-established by the investiture of one of the most influential of their number, and I apprehend no further interruptions to our operations.'[52] The Empire followed the telegraph as much as the telegraph followed the British flag. However, issues of sovereignty, penetration, and control, did not just occupy European and British Indian officials but were reflected throughout at different levels of the process of installation and maintenance of the telegraph.

A brief catalogue of the difficulties, the collaborations and resistances, facing the team shows how aware the locals were of the importance and meaning of the apparently innocuous telegraph. While rulers were quick to agree to its construction in order to exhibit their interest in modernity and technology, as in the Ottoman case, in reality their aim was to extend physical control over territories and peoples they had little power over except in terms of some hazy claim of sovereignty. As Goldsmid reminded himself, he had to 'negotiate with Persia as well as the Afghan and Baluchi chiefs over whom Persia claims a token sovereignty'.[53] East and West Mekran only needed a frontier to form what, according to Goldsmid, were actually two distinct halves: the former, with the exception of Gwadar on the coast, which was under the Imam of Muscat, was under the rule of the Khan of Khalat; the latter was under a number of petty chiefdoms. The imperial 'gaze' was turned to hitherto unseen areas and waters; Goldsmid recorded how his attempt to recruit a 'native pilot with some knowledge of local soundings' failed because the 'inhabitants were unaccustomed to such requisitions, and admitted possession of no such qualifications for compliance.'[54] The processes of imperial surveying, mapping, sounding, classifying, and transfixing were in clear display as these early lines were built taming, naming, and transforming an alien landscape and its people into something familiar, known, controlled, and judged.[55] Goldsmid recommended that under 'no condition, so far as British interests are concerned, to suggest interference [armed?] in Seistan.'

With regard to the Mekran frontier he urged the Government of India to enter into a treaty with the Khan of Khalat and negotiate on the Khan's behalf Khalat's western boundary with Persia. Goldsmid cautioned that 'the problem [with this course of action] is that the Baluchi chiefs west of this line will object to being given over to Persia.' However, the question had to be addressed 'if we are to run the telegraph line through their country.' He suggested a stratagem of informal control through bribes:

> We may come to an understanding with the Persians to the effect that we are to subsidise the Baluchi chiefs up to the western limits of

the territory formed by the Imam of Muscat under treaty, as though they were independent, exercising no interference in their suzerainty, in which we neither acknowledge nor disavow the Persian claim. Where tribute had been formerly paid, as along the coast west of Gwadar, we might suggest that an increase be accepted in consideration of the sum paid to Baluchi chiefs for the protection of the line. East and West Mekran: which only need a frontier to form what are actually the two distinct halves.[56]

An experienced officer sent warnings to the British Consul at Aleppo that he did not doubt that every bit of wire within the reach of these tribes would be stripped off the telegraph poles to make stirrups for their horses.[57] The two main concerns were, first, the best means of dealing with the tribes on the outskirts of Mesopotamia, and, second, the best way of extending the telegraph below Baghdad. A landline joined Karachi with Gwadar, the farthest point in the west that had been reached by the Indian telegraph.

Underlining the political nature of telegraphy and the questions of sovereignty that it raised, in 1863 the Governor of Banpur warned that he would attack if any attempt was made by the telegraph engineers to cross into his territory, which began at Gwadar.[58] However, he was willing to allow construction if written permission was obtained from the Persian government.[59] Even when permission was obtained from the authorities local conditions and the warrior tribes saw to it that the sanction served as no guarantee.[60] The line connecting Musandam and Gwadar, and the line from Shiraz towards Teheran was interrupted several times.[61] There was also the destruction of the line between Mosul and Baghdad and negotiations that had to be carried out with the Shamnan Bedouins. However, officers actually carrying out the survey and tentative construction of the lines believed that there was no widespread animosity towards the telegraph and the disruptions that occurred were often the result of individual or group mischief rather than any systematic endeavour.[62] This said, the attack on the workmen employed on the landline re-emphasised the need to move along the coast and return to the safety of submarine telegraphy. Afghanistan continued to be a problem for British diplomacy and control and its need to control the land routes to India.

Soon after landing on shore, Goldsmid found himself in a very difficult situation. A party from a nearby village approached him seeking employment, greatly incensing the workers already employed from a different village. Goldsmid got confused between the Maklabites under Sheikh Suleiman (already employed) and the Fillami (seeking employment) and ordered the Fillami to get to work. It took a whole day to resolve the ensuing uproar. He noted with chagrin, 'A small riot ensued. We have got mixed up with a wild, avaricious lot, and they did not acknowledge the authority of the Imam of Muscat.'[63] They had probably never heard of the Imam and his claim to sovereignty;[64] Arab horsemen and mobile bands cut the wire regularly and often stripped a workman on the line.[65] The favourite pastime of mounted marauders was to shoot at the white ceramic non-conductors at the top of telegraph poles since they made for prominent and interesting targets to practice ones marksmanship.[66] Goldsmid had to regularly resort to bribes and was probably the author of the elaborate system of bribes for the tribes that was maintained until the First World War. He wrote, 'A series of visitors to the *Sind* [cable steamer] demanding presents. They were sheikhs or pseudo sheikhs; at all events they managed to get some money and presents out of me'.[67] Again, 'The Arabs are troublesome visitors, but may be kept off today. They have had too many presents given them.'[68] Similarly, 'One of the Sheikhs of Khussub was presented with a pair of pistols. Maklabites and Fillamis were not happy'.[69] Sounding increasingly hysterical, he noted, 'More visitors including a sheikh of Nibur. Where is this?'[70] Finally, he panicked and records in his diary 'Threatening letter from the hill Bedouins. Messenger sent has not returned. … Arms given out this morning.'[71] The necessarily unequal distribution of arms and bribes possibly spawned a competitive race for arms and money in the region that continued into the twentieth century.[72] As Goldsmid unselfconsciously recorded, 'Covetous of everything the English intruder might possess, up to the shirt on his back, these unprepossessing natives were jealous of each other, in the matter of gifts, as dogs with a stray bone.'[73]

The telegraph employed a large number of Baluchis and Arabs and 'astonishment was expressed by the natives at the rapidity of intercommunication' but an old sheikh pointed out to Goldsmid that the local resistance was not irrational or an unscientific and primordial response to unfamiliar technology. The Sheikh explained that the wires would come first, then the traders would follow with guns, centralised control and taxation would follow, and the local people had so little to begin with. Goldsmid noted that the Governors of Yazd and Kerman would 'naturally be well pleased to keep their respective territories unaffected … by the wires, or free from immediate Head Quarter influence and centralisation.'[74] He faced more conventional resistance when the labourers went on strike, refusing to land the cable unless given special cable pay and a rupee for every man [or 2 Kirams or double pay].⁷⁵ Goldsmid noted with satisfaction that the 'Arab Wali, or Governor [of Gwadar], worthily seconded his master the Sultan of Maskat in readiness to cooperate with the peaceful enterprise of H. M. Government; that an English working office and store depot has been organised ... at Gwadar under the aegis of a Political Agent, and that so far from apprehensions of molestation being entertained by British residents, an English lady was actually living there with her husband as quietly as she might have done in Calcutta or London.'⁷⁶ In short, the telegraph was the means of ensuring central control and penetration both for the British and the various feudatories they dealt with ranging from the Ottomans to the Sultan of Muscat.⁷⁷

E. G. Browne of Pembroke College narrated how the railway line from Teheran to Shah Abdul-Azim, built with the aim of extending it south to the Persian Gulf, was 'completely wrecked' in 1888 by a mob angered by the accidental death of a man who leapt out of a running train. In the 1880s E. G. Browne was discovering the religious, historical, and intellectual worlds of the Bahai and Ezeli faiths; a pioneer in these fields of research. While Browne conceded that the mob was unreasonably incensed at an accidental death, he was not prepared to admit that the

> deep-seated prejudice against this [railway] and other European innovations which found its manifestation in this act is equally unreasonable. ... These things, so far as they are sources of wealth at all, are so, not to the Persian people, but to the Shah and his ministers on the one hand, and to the European promoters of the schemes on the other. People who reason about them in Europe too often suppose that the interests of the Shah and his subjects are identical, when they are in fact generally diametrically opposed ...⁷⁸

The Shah did not speak Persian among his own family, being a Kajar (Turk), and was detested in much of Persia as a foreigner.⁷⁹ There was no shortage of collaborative offers once the propaganda of lucrative Empire and technology had seeped through; 'Why don't you take Persia ... you [British] could easily if you liked', a Persian promised Browne.⁸⁰ Britain, committed to the upkeep of the telegraph empire, represented herself as the rational, good, scientific and altruistic ruler: an apparently unwitting victim of the bargain basements of Asia as much as she created their sale. It is important to note the nature of the sub-imperialism of telegraphy and its potential for such a display of power. The massacre of Bengalis employed on the telegraph, postal, and clerical establishments during 1857, the penetration of Burma and Persia by faceless Indians who formed the vanguard of the Empire up to, at least, the First World War,[81] the virulence of puppet regimes, Native States, semi-independent principalities, more vicious than the Empire itself, reflected the subservient and sub-imperialist roles allowed the members of an informal Empire.

Before moving on to the global and Imperial stage it is relevant here to comment on the 'Telegraph Princes' of the Ottoman and Persian Empires. The charge of the telegraph office was an important political function and royal relatives were appointed. The posts were also sold to the highest briber at a very high price. Goldsmid, surveying the Indo-Turkish landlines, noted that one Sulaiman Bey, Superintendent of the Telegraph, was 'said to have put some hundreds of prisoners into the telegraph office lower rooms; hardly a purpose intended for them [the office]. Not a popular but obviously a powerful man.'[82] In another instance, the Prince Telegraph Superintendent at Qom earned much more than the £4 that was his salary because persons in charge of the telegraph offices were a 'power in the land' and by accepting bribes from everyone, he maintained the 'proper neutrality befitting a Civil Servant.'[83]

A world of information and telegraphy

Newspapers claimed that 'the telegraph clerks, scattered ... over all the eastern world, are the eyes and ears of the British people.'[84] Telegraph offices and personnel were an important source of intelligence; transmitting news and information, they also acted as watch posts of the empire. When Browne visited Persia he followed the telegraph lines and enjoyed the hospitality of telegraph officials on the Indo-European Telegraph Department and the Persian lines. These officials ranged from military and civilian Europeans to Armenians, Eurasians, Parsis, and Persians. In contrast, Colonel Stewart, on his mission, scrupulously avoided telegraph stations since he was 'aware that, in Persia, a weekly dispatch is sent from all telegraph stations to the Minister of the Telegraphs for the information of the Shah, describing any remarkable person who may have passed through the village during the week.'[85]

One of the main projects of Stewart's journey through northern Persia disguised as an Armenian trader called Khwajah Ibrahim, from Calcutta was the survey of Russian oil refining and extracting activities at Baku. Stewart had to avoid British Consular Agents such as Mirza Abbas Khan at Mashhad, a 'sharp' man, since the Mirza would report Stewart's arrival to the British Minister at Teheran, Sir Ronald Thomson, who would immediately ask for Stewart's return because his presence in the Perso-Turcoman borderlands without official permission was illegal. If Sir Ronald did not issue orders immediately, his Russian counterpart in Teheran would insist that he did so. Colonel Stewart confessed that, in fact, 'I was in disguise more to hide myself from the British and Russian Ministers than from anybody else.'[86] After travelling through Persia in disguise from 29 September 1880, he finally announced his arrival to the British Consular Agent at Mashhad on 9 February 1881.

A few days later, Captain Gill of the Royal Engineers arrived in Mashhad, and Mirza Abbas Khan promptly telegraphed Sir Ronald, British Minister at Teheran, who ordered Gill's immediate return to Teheran.[87] Colonel Stewart also met the Governor General of Khorasan, the Shah's brother, and knowing that he employed an elaborate network of spies, asked him whether he, the Governor, had ever heard of Stewart's presence on the northern frontier. Apparently, only once had Stewart been reported close to Mashhad, after which the trace was lost. A man spotted him surveying and reported back to the Governor that he had seen an Armenian surveying. The Governor realised that he could not be an Armenian but must be a European surveyor.[88] The Persians were not the sole intelligence gatherers and Mirza Abbas regularly sent a man to the Russian camp at Geok Tepe to obtain information about events during the siege by the Russians. The Russians besieged Geok Tepe between 1879 and 1881, achieving their objective after three years. Stewart used to share a smoke with this man and get information about the siege. Mr O'Donovan, correspondent for the *Daily News*, was stationed at Daragaz to supply the world with the latest news.[89] It is clear that behind the apparently remote and untamed image of northern Persia there lay a bustling web of information networks, intelligence gathering and media activity conducted through the telegraph; the Great Game continued.

The telegraph as imperial context

The telegraph also provided the means to renew the magic of Empire. Exploration, discovery, and journeys across remote frontiers spawned the second generation of romantics and heroes of the British Empire.[90] Persia and Turkestan became popular wandering grounds for writers of exploration fiction.[91] Tourists complained about the characterless telegraph rest houses and preferred Persian caravanserais.[92] In the 1860s in Persia, Goldsmid was defining political boundaries, Lieutenant Stiffe was surveying the uncharted waters, F. C. Webb, the most senior of

Sir Charles Bright's engineering staff, was fixing and locating faults on the cable and gaining invaluable experience of submarine cable laying, while Lieutenant Colonel Patrick Stewart, Royal Engineers, in-charge of the Persian Gulf operations, was wasting away from an undiagnosed illness that would eventually claim his life in January 1865 in Constantinople, aged 32. Stewart left behind a widow who had only briefly known him in their four years of marriage, most of which he had spent building and managing telegraph lines in India, on the cable from Gibraltar to Malta, the Persian Gulf project and finally the Indo-Turkish lines, and he was scheduled to oversee the line towards China. Telegraphic communication between India and Britain was achieved only a few months after his death.[93] Browne, exploring the intellectual horizons of Islam and its variants in Persia, recorded his feelings upon receiving a telegram from Cambridge, translated into Persian:

'Khwáhish dáram idhn bi-dihíd shumá-rá barái mu 'allimí-I-farsi taklif kunam. Níl' ['Please authorise name candidate for Persian readership, Neil.']. I was rather overwhelmed by the reflection that even here at Kirmán I was not beyond the reach of that *irrepressible nuisance of this age of ours, electricity.* (italics mine)[94]

Colonel C. E. Stewart was an equally distinguished pioneer and behind his reticent and self-effacing prose was one of the early prospectors and advocates for petroleum. He first visited the Baku oilfields in 1866 when there were only two refineries but when he revisited Baku in 1880 there were over a hundred and it had emerged as the headquarters of the Russian petroleum industry. He warned that the Russians had progressed beyond the experimental stage to use petroleum to power their ships and railways. A Mr Nobel ran one of the largest refineries at Baku. Comparing American and Russian oil production and the emergence of spin-off chemical industries producing Naphthalene, Benzol, etc., Stewart urged that the future lay with petroleum and not with coal. He argued that petroleum was the means to switch from a slow coal-consuming navy to a fast, fuel-guzzling navy. He argued that Britain should not lag behind Russia and America, and that Britain should urgently overcome the 'great ignorance' that prevailed over the uses and application of petroleum and increase investigations into the subject. Petroleum deposits discovered in Upper Burma, on the western coast of the Red Sea, in India and Australia, meant that Britain need not depend wholly on the Russian or American sources.[95] Browne's close proximity to the telegraph lines and Stewart's scrupulous avoidance of the telegraph while in disguise, illustrate in two different ways that the informal empire of telegraphy was already in place.[96] Stewart met General Schindler, Inspector General of the Persian Telegraphs, prospecting for gold at Sultaneh to clear up the issue of hidden gold mines at the Shah's request.[97]

By the end of the century, complaints were made in the opposite direction to Goldsmid, disclosing the fact that political functions swallowed the duties of the telegraph officer. Robert Campbell wrote to his 'beloved Amroo' that he was quite safe with 250 soldiers and two machine guns under his command and that he was 'called away to take messages (confidential) ... the whole substance of which is that I continue Political work...'[98] He noted, in frontier conflict with the Baluchis, 'the traders and suppliers at Gwadar, and Jask have my orders not to supply any provisions without my written orders' and concluded that '[I]f I am while in charge of this section to be also Political Officer of the coastline, the Government must allow me to deal with the Baluchis as I suggest or redeem me of the duties.' He was subsequently offered a consular appointment in the Persian service but turned it down because of 'close study of international law, of Arabic and Persian, learning duties of the appointment, separation from Girlies (Amroo, Bun and Bell) for an indefinite period of time for retirement too far off', and, the pay was roughly the same.[99] Similarly, Electrician to the Department, Sir Henry Mance, worked with signallers on the Frontier Boundary Commission.[100] This counterpoise, both in terms of prose and function to Goldsmid, less than 40 years later, highlights the political and penetrative nature of the telegraph; it served both as a site of defence and an instrument for the empire. It modelled the language and function of the individual as much as the system. The telegraph subsumed Goldsmid's political task in the 1860s and 1870s, while politics overshadowed Campbell's role on the frontiers of empire in the 1890s. The recurrent inflammations on the nodes and frontiers of empire were not a feature exclusive to the functions of expatriates and overseas capital investment involved in the context of their temporary location, but, because they depended on the telegraphic transmission of information on gambles on the 'annihilation of time and space' in a system of circulation of promissory return on the future consumption of goods and services. Turbulence surrounding the men stationed at the frontiers of the empire was not a matter of surprise; they were not only agents of empire but also instigators of empire leading lives on the edge and capable of controlling the transmission of information.

A similar if harsher process was deployed in Upper Burma. Telegraph stores were contracted for as early as 1853 for proposed lines through

Burma. Writing from Mandalay, Sir W. M. N. Young of the Indian Telegraph Department (1871–97) recorded the problem of 'dakaits' and the turbulent state of Ava in 1888. The telegraph working party under Young was attacked and a sepoy killed. Demonstrating the political nature of the telegraph, he described the facilities arranged for his protection:

> I have an escort of 30 rifles ... Bodyguard of 4 mounted police, who accompany me everywhere ... I have a revolver and at night have it under my pillow with a Winchester repeating carbine, 14 shot, at the head of my bed. ... I have 3 sentries at night around my camp, who challenge each other at regular intervals ...[101]

The Director General echoed Young but reported that 41 offices and 1400 miles of wire had been laid during 1877–8 in Upper Burma as replacement for those lines constructed immediately after occupation and subsequently completely destroyed, illustrating the savage resistance the telegraph personnel encountered in their imperial mission.[102] The turbulence at frontiers of empire and the loneliness of telegraph stations and signallers posted in the middle of nowhere generated both humorous and tragic incidents. Stories such as the one of the drunk signaller leaving his donkey in charge of the telegraph office while he went about his business contrast with the insanity of the signaller who went on a rampage shooting innocents until he was shot himself.[103] Whole generations became telegraph men, especially in the Indo-European Telegraph Department. For example, E. M. Norris was born at Jask in 1875 where his father was working for the department and joined the department like his father.[104]

The telegraph empire was the context within which fresh annexations and zones of influence were carved during the period after 1870. The immediate focus of the telegraphic penetration moved away to the Atlantic Ocean, Africa, Australia and South East Asia. British India became a base for the launch of several telegraph schemes and imperialist penetrations. However, this does not mean a mere revival of what had been before a zone of cultural exchange but a zone of influence and control backed by the armed might of British India. The new imperialism did not acquire inhospitable lands because of political imperatives of the frontier but because of the need to secure its lines of telegraph communication. Cecil Rhodes planned to form a British South Africa Company along the lines of the Imperial East Africa Company and was granted a Royal Charter in 1889, which, while banning 'monopolies of trade' gave the railways and telegraphs to him as communication monopolies. Rhodes claimed that his 'ultimate objective is to connect with Telegraph System now existing in Egypt.' The anxiety about the corridor to the north of Africa and Rhodes's project to take the rail and telegraph to the river Zambezi and control Bechuanaland, Matabeleland, and Mashonaland shows how communication and access, especially railway and telegraphic, were an important strategic concern of the British in Africa. Rhodes wrote confidentially to the British Government, offering to extend the telegraph from Salisbury, in Mashonaland, to Uganda without any contribution from government. He declared the 'railway is my right hand and the telegraph my voice.'[105] Thus, at the local level it depended on aggressive, penetrative, and transformative action. At a physical level a place like Gwadar with 200 houses and 1000 inhabitants, and a small fort belonging to the Imam of Muscat became transformed into a vital strategic element of British control and interest over the southern coast of Persia.[106] Telegraph stations provided sources of local employment, wealth and prosperity. At the same time, it destroyed traditional practices: while Telegraph Superintendent at Dahana Baghi, Norris was approached by a Baluchi who demanded compensation for the telegraph land they had previously cultivated with cash crops and had maintained a water channel for its irrigation.[107]

The argument that the British Empire expanded until it reached either natural boundaries or met with a power of sufficient strength needs to be qualified. Certainly, ancient trade routes and networks testify to Indian trading and religious presence in remote parts of Persia and Afghanistan, and the still vital networks connecting these regions to the Indian heartland in the 1860s and 1870s testify to the expansionism of trade, but, perhaps, fundamentally different in nature from the armed and inherently monopolistic 'free trade' of the Empire. This section argues that the idea of imperialism impartially administering free trade is not completely true and statements such as 'official demands on behalf of British interests overseas never went beyond equal favour and open competition; non-intervention in the internal affairs of foreign states was one of the most respected principles of British diplomacy'[108] contrast with the reality of oppression and economic manipulation at the periphery. The empire of so-called free trade was possibly one of the most brutal and blind regimes of control when seen in terms of exploitation of privilege and monopoly, and the emergence of cartels after 1900.

Goldsmid recorded the presence of Armenian traders operating between Muscat and Bombay; they took wool and pressed it at Bombay where their agent was Dosabhoye Mirwanji, possibly a Parsi.[109] *Basrai* pearls were a prized commodity in India, especially under the Mughals. There were the Daudis at Bideshk (interestingly, 'bidesh' is the Bengali/Sanskrit word for foreign land) who practised a religion close to the original Buddhism and different from the Burmese or Ceylonese variants.[110] At Baku were the remains of the 'Temple of the Everlasting Fire'; an allegedly everlasting fire tended by Hindu [Sikh] priests, a twin of the Jwala Mukhi in Kangra District of Himachal Pradesh in India,[111] and Baku was the past home to pilgrimages.[112] Baku also had the remains of a temple dedicated to Shiva. A photograph of an inscription in the temple reminds one of Nagari scripts, and a dedication to Ganapati. There was a Punjabi priest living there who tended the fire: he had previously worked as a priest at the temple at Kangra. The author was certain, as was commonly believed by European travellers, that it was not a Zoroastrian fire temple.[113] It is equally crucial to note that this was the site of the later Baku oil fields: fissures on the surface of the earth, because of emissions of natural gas, kept the flame going. One of the main objectives of Stewart's journey through northern Persia was prospecting for petroleum and surveying Russian control over the Baku oilfields.[114]

Imperial penetration through the telegraph opened up new routes, centres, and means of communication but many earlier routes and connections were severed as territories and movement became more strictly defined and contained, and expansion and maintenance of the informal empire in this region often went hand in hand with a more violent, repressive, orthodox and centralised Arab, Persian, and Turkish Islamic sub-imperialism at the cost of smaller communities such as Hindus, Sikhs, Armenians and Parsis.[115] In contrast, attempts to protect members of the empire often led to an engagement with local powers and the reassertion of the final authority of the British. When Stewart revisited the Fire Temple at Baku in 1881 the last priest had returned to India, and surrounded by the derricks and refineries of the oilfields, the 'eternal fire' was extinguished forever.[116]

Conclusion

Telegraphy attracted power and centralisation and became a key dimension of Britain's informal empire. Imperial expansion and informal penetration occurred at several levels. First, India became the launching pad for British influence in the Indian Ocean, western Asia, south East Asia, and East Africa and South Africa. The Indian army backed up this influence and the ideology. Second, this informal involvement generated a fragile system of treaties, bribes and alliances in these regions, which saw increasing direct imperial involvement. Third, this period of expansion of the overland system was replaced by a period of submarine expansion discussed in the next chapter. Fourth, the telegraph allowed centralisation in west Asia and Ottoman and Persian sub-imperialism. This is a very significant element of overland and submarine telegraphy illustrating how telegraph technology is welded to power and hierarchy, producing local sub-imperialism and reinforcing centralising hierarchies. Finally, the telegraph provided the context from which fresh interests and discoveries such as petroleum or Bahaism emerged. Telegraphy created nodes of communication, but these also became sites of conflict and contest between local people and imperial and totalitarian interests.

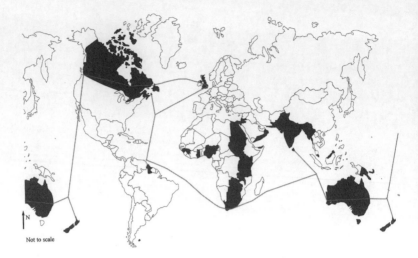

Map 5.1 The Imperial all red line with the British Empire shown in black

5
The Magical Mystery Tour: Cable Telegraphy

Submarine telegraphy is one of the greatest achievements of the Victorian or any other age.
Lord Mayor of London, 1894[1]

The demonstrator turned suddenly to Charlie, and said impressively, 'You are about to be initiated into the awful mysteries of cable making. Are you prepared?'
'I am,' responded Charlie.
'Then let us enter the halls of modern magic,' rejoined his companion ...[2]

Introduction

The period 1860–80 saw the zenith of cable telegraph technology as the first global electronic communications network was successfully installed. India was at the centre of this network that linked the imperial powers of the west to the colonies and captive markets in the east. The Indian subcontinent was the hub that controlled these lines that stretched from western India to Europe and eastern and southern India to south East Asia and Australasia. It was the crucial element in the imperial telegraph system. The period 1860–80 was one of immense expansion, frantic building and the return of the romance of travel and exploration through the construction of telegraph lines. It saw an enormous investment in terms of money and men, as the project to girdle the globe with lines of telegraph was taken up with fervent zeal. Below the surface of this picture lie relatively uncharted tales of contesting technologies, technological fashions, standardisation of speed and cost, and competition between various telegraph companies, nations and men.

The silver jubilee of the corporate giants Eastern and Eastern Extension Companies was celebrated with magnificent pomp at the Imperial Institute in London in 1894. Between 9000 and 10,000 invitations were issued and over 5000 people attended. The party ended at two in the morning. It was an apt combination: the Imperial Institute and the future monarch at the heart of the imperial system. The Prince of Wales sat at the centre of an enormous telegraphic web of over 152,000 miles, and sent off telegrams to all corners of the world. He received replies very quickly and apparently enjoyed himself thoroughly. The list of those who sent the Prince of Wales congratulatory telegrams included the Viceroy of India; governors of New South Wales, Victoria, Tasmania, New Zealand, Queensland, West Australia, Hong Kong, Singapore, and Natal; the Cape High Commissioner; the Canadian Governor General; ministers from Peru and Chile; and the Commissioner of Bulawayo. The banquet was a glittering affair presided over by the chairman of the companies, Sir John Pender. Nothing, it seemed, could stem this tide of imperial peace, progress and prosperity under the system of telegraphy.

There was talk in the air of connecting America and Australasia by a cable through the Pacific Ocean, thus completing the encirclement of the globe. This was the steel girdle that symbolised and crowned the scientific empire. 'Armed science' was the phrase used at the banquet by the very distinguished scientist Sir William Thomson, later Lord Kelvin. It seemed that the electrical sun would never set on a British empire based on the telegraph system.[3] However, a little more than 20 years later, the world was on the brink of the First World War, and wireless and telephone stations were mapping a new electronic communications network. The simultaneity of these processes and events is arresting. It was symptomatic of the demise of a world order based on a system of technology that after 1900 could foresee its own death.

It must be remembered that there were a mere handful of cable companies in the world, largely controlled and financed by Britain. In the 1894 silver jubilee banquet, amidst the glittering lights, most speeches and newspaper articles echoed the rhetoric of imperial progress. The *British Australasian* wrote on 26 July 1894:

> The Prince of Wales, standing as it were at the very hub of the British dominions, dispatched simultaneous messages by cable to the uttermost corners of the earth where the British flag flies, and in a few minutes received back words of greeting in reply. ... No factor conducing to the maintenance of the Empire is more powerful than the submarine cable .... its prompt utilisation prevents that friction

between far distant parts of the Empire which, were no rapid interchange of explanations possible, might develop into misunderstanding and alienation. The chances of the cohesion of the British Empire have been vastly increased by the progress of ocean telegraphy all over the world. The cable counts for peace and diplomacy works along its strands.[4]

Metropolitan discipline and control was celebrated through this rhetoric. Sir John Pender claimed that the 'telegraphs know no politics.'[5] This imagined world of the telegraph controlled by British metropolitan laboratories that engineered this vast empire of science: Armed Science' was Lord Kelvin's evocative phrase.

The Persian Gulf cable and the return of submarine telegraphy via the metropolis

A submarine cable linked Gwadar and Fao in 1862 after an extensive survey of the Persian Gulf carried out by Captain A. W. Stiffe of the Bombay Marines.[6] In 1864, a submarine cable linked Karachi and Gwadar to further strengthen the link. On 27 January 1864, with a line connecting Fao and Baghdad, Britain and India were linked by telegraph. In the process, Gwadar was changed from 'an obscure fishing village' on the Mekran coast, and Fao from a 'very molecule amid hamlets',[7] into two of the most important nodes in the overland imperial communications chain from Europe to India. This transformation of space and geography illustrates the complex re-mapping carried through by imperial communication systems.

This was in some ways no different to the shifting of capitals and the redistribution of land rights of the past. What was crucially important was that the court no longer travelled with the king. Often there were no stations of power; the telegraph merely travelled through. The Anglo-Turkish Convention of 1864 proved invaluable in securing Turkey within the imperial communication network and for ensuring that there were special lines reserved for traffic from India to Britain. Dawud Effendi, Director General of Telegraphs in Turkey, was extremely efficient in constructing and maintaining the lines. The strategy adopted was to employ Arab horsemen and personnel for protecting and building the lines. This provided local support for telegraph lines west from Baghdad.[8] A significant number of extra-metropolitan personnel contributed towards the success of the telegraph enterprise: Anglo-Indians, Eurasians, Indians, Arab tribes, Persian elites, Turkish officers, Europeans, middle and working class British all contributed at various points to the success of the telegraph enterprise. A number of senior officers and engineers of the Indian and Indo-European Telegraph departments joined the cable companies in senior positions. At a different level, experiments with rates on traffic and telegraph codes were first tried out in India along with the first trial of Duplex telegraphy over long distances. Knowledge generation and technology transfer was not uni-directional as seems to be suggested by the static models of Headrick[9] and Basalla,[10] but flowed both ways.

William Thomson's design and instructions were meticulously followed. William Thomson was born on 26 June 1824 in Belfast, Ireland, and died on 17 December 1907 in Netherhall, Ayrshire, Scotland. James Thomson, his father, was Professor of Engineering in Belfast and when William was eight years old, his father was appointed to the chair of mathematics at the University of Glasgow. William attended Glasgow University from the age of ten; at that time, universities in Scotland competed with schools for the most able junior pupils. Thomson began university-level work in 1838 when he was 14, and studied astronomy and chemistry. The following year he took natural philosophy courses, which included the study of heat, electricity and magnetism. In 1841 Thomson entered Cambridge and an important paper on the uniform motion of heat and its connection with the mathematical theory of electricity was published in 1842 while Thomson was studying for the mathematical tripos at Cambridge. Thomson was unanimously elected Professor of Natural Philosophy at Glasgow University in 1846. He became one of the experts on submarine telegraphy and dominated the science and engineering of much of the Victorian period.[11] Thomson was knighted in 1866 and made Baron Kelvin of Largs in 1892. He died in 1907 and is buried in Westminster Abbey in London.[12]

The success of the Persian Gulf cable renewed interest in the Atlantic connection. The use of O'Shaughnessy's pioneering efforts in crossing rivers in India proved valuable for the subsequent perfection of cable technology.[13] Thomson's position as the premier scientific expert on submarine telegraphy was also established through the success of his experiments in cable technology. After this, the great laboratories at the University of Glasgow and the Cavendish Laboratory at the University of Cambridge, controlled the legitimisation of research in telegraphy. Thomson's expertise was contained within an institutional space and legitimised by a variety of social forces.[14] Science, in this period, was enclosed within the laboratory, university, and technopole, as much involved with research as with the standardisation and closure of technological experiment.[15] Implementation and usage was carried out by different groups of people in other places. For example, Siemens and Halske in Berlin were acknowledged to be the best manufacturers of the instruments designed by Morse in the 1850s.[16]

In fundamental ways, the language of metropolitan control and generation did not match reality. Schaffer, in an unpublished paper, demonstrates how the metropolitan claim to universal applicability had to be repeatedly proved in the outposts of the empire. This book goes beyond the idea of empire as context to argue that the laboratory was in the colony in the case of the telegraphs; extra-metropolitan experience transcended the narrow confines of the laboratory to the extent that the submarine telegraphs were not only born out of extra-metropolitan but extra-laboratory experience.[17] For example, William Thompson's expertise combined with the local knowledge of Lieutenant Colonel Patrick Stewart, and A. W. Stiffe, engineers and the workers of the Indian Telegraph Department, and the Persian overseers and guides, to build the Persian Gulf Cable that produced the knowledge and confidence to rebuild the Atlantic Telegraph Cable.[18] Another experience was the first cable in 1861 between Malta and Alexandria via Tripoli and Bengasi, laid along the African coast by the British government. The success of the Persian Gulf cable allowed the British to translate their experience from other parts of the world into the transatlantic cable.[19] There are three important points to be made here. First, a substantial number of people involved in the Atlantic and the previously implemented lines to Asia and Africa were the same, ranging from O'Shaughnessy, Morse and Thomson, who travelled on the same ship to try and work out a way to connect US with Britain to A. W. Stiffe, responsible for repairs and maintenance of the Persian Gulf cable. Second, this created a pool of people whose experience and interest started with practical problem solving. Third, this experience and practical problem solving was based on experience outside the laboratory and the metropolis.[20]

The *Report of the Joint Committee on Submarine Telegraphy* submitted in 1861 represented the synthesis of all that was known about the technology up to that time.[21] Captain Galton, Royal Engineers, investigated the Red Sea telegraph cable. Four men from the government and four men from the Atlantic Cable Company compiled the report, that Cyrus W. Field, the American entrepreneur behind the Atlantic cable, noted was 'of great value in forwarding the cause of ocean telegraphy.' Much of this knowledge and experience was extra-metropolitan in origin. Thus the laboratory, in this case, had moved beyond the confines of the narrowly metropolitan and institutional. However, the main reports such as the one presented before the parliament did not mention the extra-laboratory and the extra-metropolitan experience.[22] Two things were happening: first, Asian and African experiences of the pool of experts were underplayed; second, and more crucially, experience itself was undercut and the sanitised laboratory in metropolitan settings privileged.

C. W. Siemens clashed with C. F. Varley, C. V. Walker, and other British telegraph experts over two sessions of the Society of Telegraph Engineers after the reading of his brother, Dr Werner Siemens. The claims were: first, that the first successful experiments using gutta percha for cables occurred in Prussia; second, Siemens was the author of the equation for proper laying of cables and the detection of faults in them.[23] Varley and other experts disputed both claims and pointed out that 'Submarine cables are undoubtedly of English origin and design' and that Britain built the machinery to detect faults; all Dr Siemens had done is provide the formula. In response, siemens claimed that while 'every submarine cable, which is now working is, almost without exception, the produce of this country [Britain], and has been shipped from the Thames' and 'although submarine telegraphy is undoubtedly a English enterprise it must be admitted that much has also been effected abroad to bring appliances to their present state of perfection.'[24] After the first successful Atlantic cable, shares of the cable companies increased rapidly in value. Huge, perhaps disproportionately so, monies were raised and submarine telegraphy became the most popular technology. The problem of closure was that it was a technological process that strove towards homogenisation and standardisation as much as it engineered its replacement. Standardisation and replacement seem to have been enmeshed in the case of the telegraph.

The 'girdle around the globe': The imperialism of the cable companies

Ironically, while the British Post Office was nationalising the telegraph in Britain, the field of submarine telegraphy was left almost entirely to private, British hands. Gladstonian Liberalism, contradictorily, saw both an engagement with centralisation and state control in Britain, and an apparent disengagement with the empire and the establishment of private interest abroad. The issue of imperial federation with central control versus Gladstone's vision of self-reliance and voluntarism was reflected in the local arena of British parliamentary politics. John Pender lost the election in 1868 but won as a Liberal candidate for parliament in 1872, and held his seat up to 1885.[25] This period was also the zenith of the cable empire without the shadow of a cloud on the horizon. Pender, having lost his seat to the future Liberal Prime Minister Henry Campbell-Bannerman in 1885, remained in the political wilderness until 1892, when he was re-elected to parliament. Pender envisaged a telegraphic empire under a homogenous system and one that was controlled centrally by a few. Pender, an opponent of Gladstone's Home Rule Bill for Ireland, was a Liberal-Unionist and believed that Ireland should forever remain united with England.[26]

When the submarine telegraph cable boom started after 1864 there seemed to be no end to the potential for the progress of telegraphy. In 1864 there were 1100 miles of submarine cable in the world. By 1870 the figure had jumped to 15,000. In 1900 there were over 210,000 miles of cable, excluding overland telegraphs.[27] Historians such as Baark and Ahvenainen have dealt in detail with the Caribbean and Chinese telegraphs,[28] Headrick, among others, has dealt with the global picture.[29] However, before turning to the formal and semi-formal imperialistic telegraph moves over land it is important to consider a series of events that made a commercial group virtually a partner of Britain in the British Empire.

Submarine telegraphy, enriched by its extra-European and extra-metropolitan experience, embarked upon another venture to join Britain and America across the Atlantic Ocean. In 1865 the US and Britain were joined by cable. John Pender and Cyrus Field were the two main entrepreneurs behind the enterprise of submarine cables joining Britain and the US. Submarine telegraphy, financed by private capital, occupied a popular share of speculation and profit after the success of the Atlantic cable.[30] For almost 30 years the cable empire of the Eastern and Atlantic Telegraph Companies, under different guises, dominated the telegraphic world and enjoyed satisfactory profits for more than a decade.

A series of experiments, innovations and a wide variety of skills went into the manufacture of the cable empire. William Thomson's Mirror Galvanometer, to be replaced by the Siphon Recorder, was specially designed for long cables because the Morse code was distorted when transmitted over 700 miles. Rather than wait for the entire clearing of the line before a fresh signal could be sent, it depended on changes in current strength. The concentration of expertise in the Report of the Submarine Telegraph Committee to parliament was complemented by the *Telegraphic Journal*, later the *Electrical Review*, founded by John Pender in 1861 to collate in the public domain all that was known about the telegraphs and electrical research.[31] Cable construction improved after Pender merged the Gutta Percha Company, which had a virtual monopoly over the supply of gutta-percha, and one the premier cable manufacturing companies, Messrs Glass Elliot, to create the future giant Telegraph Construction and Maintenance Company (Telcon) in 1864. Porthcurno became the centre of cable training and signalling practice. Telcon manufactured two-thirds of the cables used throughout the world before 1900.[32] By 1880 the production of gutta-percha became virtually the monopoly of the India Rubber, Gutta Percha and Telegraph Works Company supplying Telcon, its sister concern under Pender's chairmanship.[33]

Writing in 1872, a Public Accountant and Auditor of the Marseilles, Algiers, and the Malta Telegraph Company, noted that seven companies were involved in the submarine system linking Britain, India, and the East, and that the cables belonging to them were all manufactured and laid by the Telcon. He noted that it 'should be understood that all these companies were brought in connection with one another, and not only is the traffic managed by a joint committee composed of members of each of the different boards, but there is a great saving by the division of the cost of rent and expenses of joint stations, and in many other ways.'[34] The list of affiliated companies ran to 26 pages in his book beginning with the Falmouth, Gibraltar, and Malta Telegraph Company and ending with the British Australian Company.[35] Pender or his appointees were on the board of most of these companies and Pender himself was the chairman of the two corporate conglomerate giants, the Eastern Group of companies and Telcon. This emphasises the fact that this monopolistic system was highly vulnerable and extremely 'conservative' and resistant in its attitude to change.[36]

Cables were laid from Malta to Bombay passing through the Suez Canal and the Red Sea. Bombay and Madras were joined by landlines under the control of the Indian Telegraph Department. From Madras the British India Extension Telegraph Company cables crossed the Bay of Bengal to land at Penang on the west coast of Malaya and stretched south to the island of Singapore at the tip of the peninsula. Telegraph stations were opened in Singapore and Penang, and the line was opened to the public in 1871.[37] Answering the Australian demand to be linked to Britain via telegraph, Pender floated another company called the British Australian Telegraph Company, which laid a cable to Batavia (Djakarta) in Java, from where a Dutch landline led to Banjoewangie on the southeast corner of the island. The Australian government fulfilled its promise to complete a landline from Port Augusta near Adelaide to Port Arthur in the north in 1872. By the end of 1872 telegraphic connection between Porthcurno in England and Adelaide, a distance of 12,500 miles, was complete. New Zealand and other colonies were subsequently included in the system. Pender consolidated the four companies under the Eastern Telegraph Company in 1872.[38]

South America was a part of Britain's 'informal empire' and Britain persuaded Portugal to agree to the terms of Brazil's independence in 1826. Lord Palmerston spoke of over £150 million of British capital invested in South America.[39] In 1866 the River Plate Telegraph Company under Matthew Grey, the Glasgow engineer and ex-managing director of the Gutta Percha Company connected Buenos Aires, the capital of Argentina, and Montevideo, capital of Uruguay.[40] Buenos Aires was joined to Valparaiso in Chile by 1872. Portugal and Britain were connected with Brazil in 1873. The process of telegraphic connection through South America continued while Pender's Cable Empire swallowed the lines of its smaller rivals such as the River Plate Telegraph Company.[41]

The emergence of cartels and monopolies

Reflecting the way in which cable cartels controlled the flow of information to the east, a dual monopoly emerged in North America with consequences for the circulation of information in the US.[42] In 1866, the Western Union emerged as a national telegraph monopoly that controlled the channels of information to the New York and Western Associated Press; the purveyor of 'news' to papers in the US. The New York Associated Press combined with the Western Union in the constitution of the telegraph as a 'national network', and replaced the party organs and big newspapers as the primary source of national news. As large regional telegraph networks were consolidated, the Associated Press grew into an organisation that collated all information in New York and sent consolidated reports to all newspapers across the country.[43] The nature of reportage was also transformed: 'reporters were to state the bare facts without opinions.' The Associated Press employed many reporters and the editor selected the most saleable news, which was different from newspapers where editorial policy and politics influenced content and presentation. News as commodity increased in scale and beyond control as volume increased and no judgement was exercised. Blondheim, in this classic work on the flow of public information in the US, comments:

> The Associated Press monopolized telegraphic newsgathering and news distribution in America in the second half of the nineteenth century. Its structure as a national institution – impersonal, non-local, unselfconscious, and hidden – gave wire service news, however

partisan, the semblance of objectivity ... [and] helped familiarise Americans with a common information environment.[44]

Although he is enthusiastic about the telegraph transforming communication and serving to 'centralize, even nationalize, information', enervating the entire nation with one idea and one feeling,[45] Menahem shows how the extension of information monopoly and centralisation, in the short term, exaggerated difference and dissension. As the Western Union and the Associated Press emerged as the twin pillars of information circulation in the US, the nation engaged in bitter civil strife.[46] He noted that 'Americans had no inkling that an American Reuter existed' and were unaware of the monopolies over information transmission, distribution, and circulation.[47] Contradicting the initial impression of a surfeit of information, what emerges is a picture of a highly monopolised system of information transmission and distribution throughout the telegraph world. Reuters monopolised the distribution of news in the British Empire while Pender's cable empire dominated its transmission.

The failure of the sedition cases after 1905 in India was due, largely, to the fact that the government could not often legally prove in courts a source of news other than news syndicates such as Reuters ... and had to rely on issues of translation and interpretation to construct their sedition cases. At the level of capital, financial, and administrative control, and the distribution of information, the telegraph system reveals its inherent tendency towards monopoly, whether by the state or by corporations, and therefore its vulnerability to systemic crises. Arising out of the State's need for information the telegraph system was rapidly and simultaneously transformed into a purveyor of news. The demand for news increased dramatically after 1857. The telegraph used this demand to generate revenue and increase the volume of messages. From this perspective, the press was an essential by-product of the telegraph and one which would later swamp it. It would, however, be wrong to see it the other way around. The press did not create telegraphy and its expansion was not independent of it. The style and content of press reportage crucially changed because of the telegraph.[48] Events replaced descriptions, and bulletins replaced diaries in newspapers. Columns rather than pages became the favoured format. Language itself was transformed and imbued with telegraphese. Catchy headlines became the order of the day. By the 1880s and 1890s armies of correspondents were being sent out by the newspapers to explore areas and report on events. This insatiable twin-hunger, the hunger of the telegraph for messages and the hunger of the press for news, had global consequences.

Reuters, established by Israel Ben Josaphat Beer, who changed his name to Paul Julius Reuter, was one of the pioneers of syndicate journalism. By the 1860s he had a network of correspondents spread out all over the continent and the east, and his news circulated along the vast telegraph network wherever it reached.[49] He was the official news supplier for the government in India. This is not to attempt a history of journalism or of Reuters. Reuters was a symptom of the kind of news reporting that the telegraph promoted: it was about getting the most sensational news story first. Importantly, press messages were allowed a significantly lower tariff and in the 1860s and 1870s houses such as Reuters were permitted to send their press messages free of cost. These measures went a long way to creating a highly developed world of media. The role of Reuters in India is discussed in the next chapter, concentrating on the distinctions between news and information with reference to the issue of secrecy. Imperial and international news conferences promoted and fixed homogeneity and difference in the global village created by the media.

> Day and night, cables are flashing the news of one half of the earth to the other half. Indeed, were it possible to realise the comparative emptiness of the morning paper with its columns bereft of any messages from India, Australia and the East generally. … We are so accustomed to hear, the day after the bazaar has been gossiping about it, how many mango trees have been smeared…[50]

It is interesting to note that this was not the case in 1874–5 when it was noted that, although there had been an increase in number, the total was out of 'all proportion to the facilities afforded'. The total number was only 4206 whose value at ordinary rates would have been Rs. 90,273 but the actual amount charged were Rs. 17,820. That year, 7365 free news messages were sent.[51] By 1882–3 the number of press messages had reached 10,382 and paid revenue of Rs. 40,553.[52] The Press grew gradually and through substantial help from subsidised telegraph tariffs.

Associated industries: Gunrunning

The telegraph did not just mean poles and wires but also opened up new markets and routes. One immediate though relatively unnoticed phenomenon was the arms race among the Arab and Afghani tribes. From the opening of the coastal areas to sea and land traffic an enormous demand for illegal arms and ammunition began to be generated in the Persian and Afghan hinterland. The novelty of semi-automatic and repeat-firing firearms saw most of the tribes who manufactured crude firearms for local consumption switch to the arms trade. Large monies were invested in the trade and individual lots could consist of up to 30,000 weapons. Of course, the chief suppliers were the west and the colonial powers. Dumping of carbines after the South African war by colonial governments as far away as Australia and New Zealand were bought by western speculators, brought back to Britain, and then sent on to feed the illegal trade in Persia, Afghanistan, and the Northwest Frontier province, the Punjab, and Bengal in India in 1907 and 1908. Political agents managed to seize some and forwarded them to the Viceroy at Simla who, in turn, sent them back to the India office.[53] The extreme mobility of arms, men and news were direct spin-offs of the revolution in transport and communication technology.

By 1912, Viceroy Hardinge was complaining about the unremitting arms flow from Birmingham that was arming disaffected sections of the empire. Hardinge was outspoken in his criticism:

> I do not see why we should be forced to spend about £100,000 a year and incur great difficulties in the Persian Gulf merely because (our manufacturers) want too hard a bargain for their trade in arms. The first necessity is to see whether we cannot find some means of putting an end to the British traffic in arms at Maskat, which after all is worse in principle than that of the French. Is there no means of stopping the export of arms from Birmingham?[54]

It is therefore unsurprising to find bomb-making manuals and anarchist treatises printed in Paris being circulated as manuscript copies at two different ends of India: Calcutta and Pune.

The Cable Empire versus the British Empire

In the 1870s, at a conference between the Governments of New Zealand and Australia the project to lay a state-owned Pacific cable between Australia and Canada was first discussed. This was the beginning of the movement for an imperial telegraph system independent of British private capital: 'the all-red line'.[55] It was to be a 'state-owned all-British cable that could not be alienated by share transfers'.[56] Another reason for this move had to do with the high rates of the submarine companies, and the rest of the Empire asked for something equivalent to the Penny post.[57] The opponents of this scheme were the cable companies and the British Post Office, possibly under the influence of John Pender. The Colonial Jubilee Conference opened in 1877 with fresh demands for an imperial Pacific cable and demand for an all-red line. Pender offered to lower rates to Australia on the Eastern lines depicting the project as wasteful expenditure and competition.[58] Pender engineered monopolies because he believed that the Liberal ideal of serving as many people as possible through a cheap telegraph could only be achieved with greater centralisation. He resented the international telegraph conferences, which did not include representatives of the cable companies, though they were private interests whose future depended on international tariff agreements: if the empire wanted an all-red line, Pender wanted a cable system run exclusively by him and his associates.[59]

The imperialism of the telegraph is illustrated in the annexations of the mid-Ocean islands of the Pacific Ocean. Sandford Fleming, champion of the 'all red line', stressed the 'importance of acquiring such an admirably situated landing-place for the cable. ... [Necker Island] one, too, that had never yet been taken possession of by any nation, and could be had for the mere trouble of taking'. Thus, started the race for Necker Island off the coast of Hawaii, locally popular as the 'little lava rock'.[60] Britain did not wish to upset US influence in Hawaii and looked for alternative places to land the submarine cable. Inspiration for the project came from a casual conversation Fleming had with a high military official who had served in India and whom he had met while travelling. Fleming explained to him the British Government's recalcitrance over the Necker issue.

'"Ah," drily remarked the officer, "the best thing to do in a matter of that kind, is to act first, and ask for leave afterwards."' The official gave the example of the acquisition of Perim Island near the port of Aden by the British Consul. Learning from a visiting French Consul that he was on his way to claim French sovereignty over the island, the British Consul plied him with dinner and drinks while his men planted the British flag over the island and claimed possession.[61] Accordingly, Fleming enlisted a naval officer and hired a steamer.[62] Information about the plan was leaked to the Hawaiian government and amid a glare of publicity and wild media speculation, rival steamers raced for Necker.[63] The Hawaiian government was first to reach it.[64] Fleming was furious that this meant slower speeds of telegraphy and an extra cost to the Pacific cable project, something that could have been avoided 'at the mere cost of despatching a British war-ship to take formal possession of the island'. Britain annexed Fanning, Christmas, and Penrhyn Islands in 1888, 'in view of the possibility of their being utilized in connection with the projected cable.'[65] In Africa, the original plan to run an inland telegraph from the south to the north had been replaced by Pender's cables that encircled the continent with duplicate lines running down the east and west coast of Africa and connected with South America and India.[66] Politicians and world leaders clearly recognised the imperial importance of the telegraph. Accusing Lord Curzon of being an old style imperialist keen on acquiring territories, Edwin Montagu condemned imperialistic expansion for 'diplomatic, economic, strategic and *telegraphic* reasons.'[67]

There was something substantially new to the imperial expansion after 1850, and it was the telegraph, more than the railways, that provided the locus for imperialism. In this sense, this was a new empire of science and technology, and formal and informal control sprung out anxieties over routes and access. This renders the telegraph narrative simultaneously metropolitan and peripheral, economic and political, a history of science, of impact on India, and of the socio-political imagination of the telegraph network. The telegraph system was also certainly not neutral, as Pender had claimed. It was a crucial force in informal and formal empire. It did not just follow conquest. Commenting on the cable expansion it was confessed, 'in certain cases patriotism has been the main inspiration at work. In such circumstances the companies have found that trade follows the cable, with even more certainty than the flag.'[68]

By the 1880s and 1890s lower middle class and working class European, Eurasian, and British personnel dominated the telegraph service. It had become a stronghold of racially discriminatory recruitment that underlined its political function. It was stated, 'The telegraph clerks, scattered ... over all the eastern world, are the eyes and ears of the British people.'[69] To give a particular instance of their intelligence gathering, the involvement of the Indo-European telegraph department in Persia and Afghanistan was political. In the relatively unpublicised war on the arms trade in the Persian Gulf and Afghanistan, which reached a climax between 1907 and 1912, it was admitted that the staff of the Indo-European telegraph could not with safety continue their practice of surveillance and information gathering after 1910.[70] In short, the telegraph and its personnel were involved directly and indirectly with empire building and its maintenance.

The problem of tariff and traffic

In terms of financial investment British over-involvement in submarine telegraphy contributed to a sense of imperial anxiety. British long-term investment overseas had reached £60 million a year by 1870. By 1900,

Britain had £1 billion 700 million invested overseas.[71] A significantly high rate of overseas capital investment tied up over long periods of time made this structure peculiarly vulnerable. In the particular case of the submarine telegraphs, over £40 million was invested and by 1900 over 66 per cent of the cable network was British owned. If the landlines and the cables were combined, the total investment from Britain was at least £106 million by 1894.[72] Much of the capital invested with rapidity between 1865 and 1880 was private capital. This was separate from guarantees given by colonial governments of fixed annual sums of, on average, between £10,000 and £20,000 for ten to 20 years. The silver jubilee brochure of the cable conglomerate went on to note that the 'wonder of it all is that the whole system is the growth of a single generation. Within the brief period of thirty years over 150,000 miles of cable have been laid.'[73]

This was the crux of the problem. An enormous investment in technology was made, yet there was no attempt to foresee the replacement of this technology by other, more efficient means. It was admitted in 1894 that the 'great corporation ... are not merely joint-stock companies; they are also public institutions with which we could not afford to dispense with.'[74] Friends had warned Pender that he was 'throwing millions into the sea.' Perhaps, that was literally true. British desire to control the cables and have lines that were completely 'red' (the colour used for British areas on the maps of the 1890s), may have led to an over-investment in a technology, the construction of which was too rapid not to be vulnerable. Given the vast investment by the late 1890s of over £40,000,000 in the 'crimson thread of kinship' that ran through the empire, replacement by other technologies could be resisted temporarily, but not for long. A further factor was the interest of the British Post Office which threw in their institutional interest with the cable and telegraph companies. The commitment to submarine telegraphy had been very high for Britain. The 'crisis of imperialism'[75] was not just brought on by an inchoate amalgam of external and internal forces unleashed by the late nineteenth century and early twentieth century but was also, perhaps, a shock generated by the vast investments made in the telegraph industry and the growing realisation of the need to liquidate these holdings: the imperial crisis was simultaneously a telegraphic crisis.

This was a case of massive tied up capital investment. The telegraph system never yielded the fruits that were prophesied. It is important to remember that the telegraph project was a symbol of the empire and during its early period, a spectacle watched with curiosity as well as scepticism by the empire's subjects. The dilemma was simple: if the rates were low more people would correspond, but lower rates meant less revenue for the telegraphs. A sufficient increase in the volume of correspondents was never really achieved and increases in numbers were offset by lower profits. The profits made by the telegraphs sometimes barely justified their existence and, by 1910, the dream of a global free market economy with a single standard of time and the corresponding ideal of perfect exchange of information did not work in reality.[76] The easy equation between telegraphic expansion and commercial returns did not fulfil its promise. For example, the Australasian cable yielded net profits of only £980 until well into the 1880s. Experience had shown by 1894, that the Indian business was 'not an expansive one' and that the volume of traffic had 'practically remained stationary for many years.'[77] There was little justification for the system in terms of volume of business transacted. Yet obsession with control over the telegraphs saw lines being repeatedly multiplied. These were the crimson lines of kinship that tied the empire together. Sir Charles Bright was behind the West India and Panama Telegraph Company formed in London in July 1869. It had a share capital of £650,000 and, according to Bright's calculations, its annual income with subsidies was estimated at £170,000 (traffic receipts at £156,000 plus subsidies of £14,200) and the expenses at £30,000, which meant an annual profit of 20 per cent on the share capital. These expectations on the prospectus were not fulfilled: in 1871 annual takings were £10,044; in 1872 they were £26,074; and in 1873 they were £30,323. Obviously, much less than £140,000 a year and net profits in 1872 were £5101; in 1873 only £358. In short, the return on capital was less than half a per cent until 1875.[78] This enterprise, although dominated by private capital, was subsidised by most colonial governments in the form of annual subsidies.

'Progress and peace': Technological change and information panic

That there was no easy fit between market and information underlined the fact that the telegraph merely served to sharpen fluctuations and quicken the flow of short-term speculative capital invested in the prices of goods rather than the production of them. The primary dependence of the cables was not on private or commercial messages but more on the huge volume of Press messages and Government despatches. Contrary to the belief that perfect information would ensure peace and prosperity what this news hungry world did was to bring peripheral or the marginal sites of conflict into the metropolitan heartland. Blow by blow accounts by war correspondents promoted a rapidly circulating economy of information that could not be controlled by any one power but involved many of them. War was crucial to the media and therefore to the profit margins of the telegraphs. The Afghan campaigns of the 1870s and 1880s, the Boer War, the bombardment of Alexandria and subsequent operations in Egypt, campaigns in the Sudan, and Suakin, generated huge volumes of government and press messages with 'hourly news sent to Europe.'[79] It was stated to be impossible to imagine by the 1890s, '… the comparative emptiness of the morning paper with its columns bereft of any messages from India, Australia and the East generally … We are so used to hearing, the day after the bazaar has gossiped about it, how mango trees have been smeared. … Hardly a whisper can pass in the bazaar or a piece of "gup" circulate in the settlement, that the telegrapher does not it, and cannot transfer it to London if he so pleases.'[80]

This was a very vulnerable world where telegraphs, rumour, gossip and news were inextricably tied together. The telegraphic information system and its subsidiary industry, the media, depended on imperfect peace. The importance of India, due to its crucial position in the imperial network, led to the importance of Indian politics and panics on a global level. Indian unrest threatened the entire system and was therefore not peripheral. The telegraph represented the New Imperialism of the late nineteenth century and was a symbol of the empire of science that Britain claimed to represent. The decline of submarine telegraphy and the growth of a more aggressively nationalist position in India can no longer be conceptualised as unrelated spheres. The nationalists in India were quick to seize the opportunities offered by the diverse systems of communication to reach an international audience. After 1920 it was the 'imagined' state with its imperial network that had to rethink itself in the changing context of technology.

Guillelmo Marconi met with dismissal and disbelief from the cable companies, and telegraph experts during the initial stages of the wireless; newspapers accused the British Post Office of wasting public money in investigating 'a young foreigner's inventions' to the neglect of 'home research'. Contradictorily, by 1900, they were accusing the British government of having 'fought shy' of invention.[81] Marconi formed his patent company to engineer a monopoly through his wireless apparatus and code in 1907 but the British government turned him down. Perhaps, the government realised through their experience of cable telegraphy that private monopoly capital tended to maximise profits while at the same time it rapidly standardised the technology. It was very difficult for this capital to pull out of this kind of investment. Although the real breakthrough in wireless technology would come in the 1920s with the perfection of the tuning of wavelength and directional antennae, yet by 1907 the vast cable world was aware of the threat of extinction. In the middle 1860s, financiers had warned Pender that he was throwing millions of pounds into the sea. In 1894 this statement was the standing toast of the time and a celebration of the vision of a few men for the benefit of all. By 1910 an enormous investment in cables lay on the bottom of the ocean steadily becoming obsolete. The reality of unprofitability was not foreseen by the celebratory rhetoric of 1894 and this generated a social and technological crisis of far reaching consequences in the first half of the twentieth century. It was compounded when the average cost of a telegram in terms of labour and machine began to spiral downwards inexorably. Minimum profit guarantees and the telegraph monopolies tried to keep prices high, as did the various governments. However, by the end of the century, the sheer downward pressure on prices and consequently the huge growth in information transmission made the state extremely vulnerable to and very aware of the need to contain, restrict and control this volume of information slowly spinning out of control. It is therefore not unsurprising that information panics begin to proliferate through the system after the 1890s.

Imagining a nervous telegraphic world

The cable project gave rise to a vast romantic literature and the rhetorical devices underlying telegraph ideology need identification. In 1868, at another grand cable-banquet, Cyrus W. Field proposed a toast to submarine telegraphy, the 'Crowning glory and the triumph of our century.'[82] After 14 years of struggle, he asked, when 'the news reached us that the last cable had been laid ... did not every man feel that a *new world and a new time were opened to him* ?' (italics mine).[83] Gladstone sent congratulatory messages and Field toasted the 'men of science'.[84] Cable telegraphy represented, tangibly and symbolically, the ideal of scientific and rational empire; the newness of 'New Imperialism' lay not in its ideology but in its technology. William Orton, President of the Western Union Telegraph Company, wired from New York to congratulate Field on behalf of the 'countless host who, throughout the world within influence of the Press and the Telegraph, sit daily at a feast of reason, ever satisfying but never satisfied'.[85]

The telegraph system became the symbolic reflection of the human nervous system. Sir Charles Bright prophesied, '... the seas shall cover wires communicating like nerves between every great centre of thought and action in the world.'[86] Books eulogised telegraphy as the civilising and Christianising mission.[87] Poetry abounded:

> By telegraph your will has fled,
> Your thoughts made known, and clearly read;
> Informing those with whom you'd treat,
> When you started and when you'll meet.
>
> Tidings to heaven by angels borne,
> Ethereal as the light of morn;
> Proclaiming with rejoicing strains,
> Where darkness dwelt the Saviour reigns.[88]
>
> When coiled wire with silk is bound,
> The fluid keeps within its limit;
> Its track unknown within the ground,[89]
> Will find the wire in distant summit.
>
> The wires no more confined to land,
> Doth now extend across the deep;
> Join fertile shores with coral sand,
> And busy throngs with those who sleep.[90]
>
> The soul no more confined to earth,
> Soon wafts its way to endless love;
> There joins the body in its new birth,
> With grateful throngs who rest above.

A particular mix of missionary, civilising zeal combined with the celebration of science and rationalism centred on the analogy of telegraphy and the nervous system.[91] Telegraphy, especially the submarine network, was imagined as a translation between the body and the body-politic: 'They are the nervous fibres of the world, conveying intelligence from one continent to another, just as the nerves of your body send the impressions of the senses and the wishes of the brain to every limb.'[92] The liberal ideology of the free market and perfect information underlined throughout the telegraphic age, 'the importance of this [telegraphic] service to the State in relation to international affairs and its influence for good in the promotion of amity and understanding between all nations.'[93] The telegraph was a body with a brain[94] allegedly promoting a new amicability between nations. Of course, there had to be members and nations, limbs and digits.

Authors pointed out with pride that 'Submarine cables, like steamships, are, however, in the main a British invention.'[95] The telegraph rapidly became a British national concern and was heavily invested in, in terms of ideology. It was the symbol of Britain's world domination and after the mid-nineteenth century, the telegraph and electrical sciences in general represented the most prominent 'marvel' of the day. The telegraph represented to the Victorian person the apogee of achievement in his or her time.[96] It was the apex of Victorian science, 'one of the greatest achievements of the Victorian or any other age.'[97] From the 1850s huge industrial exhibitions like the Great Exhibition of the Works of Industry of All Nations celebrated Britain's scientific achievements, more sharply delineated in contrast with the display of handicrafts from the colonies. This contrast, however, did not reveal the science of telegraphy; rather, it occupied much the same place as any monumental work of art.

Writing about the Crystal Palace exhibition, the German critic Lothor Bucher wrote, 'to say that the spectacle is incomparable and fairy-like is the soberest understatement. It is like a fragment of a midsummer night's dream seen in the clear light of day.' Queen Victoria wrote in her diary entry for 7 June, 'To the Exhibition: went to the machinery part, where we remained two hours, and which is excessively interesting and instructive, and fills one with admiration for the greatness of man's mind, which can devise and carry out such wonderful inventions, contributing to the welfare and comfort of the whole world…' Dickens wrote with grumpy but uncanny foresight, 'I have always had an instinctive feeling against the Exhibition, of a faint inexplicable sort.'[98] Faced with the magic of electricity, the average Victorian viewer was no different from the reverential primitive awed by the magic and mystery of some shamanic totem, and Lord Kelvin was probably the most respected shaman of his time. It is here that the argument introduces the distinction between corporate strategy and the unpredictability of technological change. The empire was no more vulnerable to such transformation than any other corporation except, perhaps, in terms of scale. Possibly because of the large amount of private as well as governmental capital invested in it as well as the fact that this was a communication technology for public use, telegraph technology standardised rapidly to achieve status quo. Corporate strategy (I include the state within this category) tended to control and arrest technological change. In the first decades of the twentieth century, Britain could not financially or ideologically affect a rapid shift to the emerging technologies.

Telegraph eulogies noted, 'Virtually, all the energies of civilised mankind, at once in producing, manufacturing, and commercial enterprise, are regulated, often almost unconsciously, by the cable advices that are ever, without an instant's cessation, darting around the globe on the wings of electricity. ... The wonder of it all is that the whole system is the growth of a single generation.'[99] The issues concerning a new imperialism need to be further specified. The Scottish and northern British influence behind the vast telegraph enterprise cannot be overlooked. The Universities of Edinburgh and Glasgow trained many of the pioneers of telegraphy ranging from O'Shaughnessy to Thomson and Clerk Maxwell, while many of the entrepreneurs from Pender to Grey had connections with northern cities. There were two significant points that need to be recognised: first, the notion of a centralised body politic, and, second, the idea of control over change. The first saw the telegraph system in terms equivalent to the human nervous system: in a paradoxical transition the body (politic) was imagined like an electrical (electoral) machine with heads and spinal cords, ballots and networks, while the telegraph system was imagined as a biological, nervous being. The notion of translatability between states meant that this analogy could be enforced. Electricity was to serve as the means to transmute between society and machine, and as the central transformative impulse. How to preserve energy became an issue not just for scientists but also for the future of empire as it struggled with the issue of entropy. Thomson's dislike of theoretical physics reflected the Scottish Enlightenment belief in a pragmatic problem-solving approach towards issues read in primarily mechanical terms. Kelvin wrote, 'I am never content until I have constructed a mechanical model of what I am studying. If I succeed in making one, I understand; otherwise I do not.'[100] The problem-solving body was imagined to be centralised, located in a laboratory or state, and this accorded with the belief in reform and amelioration conducted by an enlightened and centralised administration. This was different from Gladstone's federalism and soft state approach towards the empire. Perhaps, the move away from landlines signalled a move towards the informal empire of the Liberal-telegraph ideology as was precariously enforced and practised by imperial Britain after 1864–5.

In terms of scientific imagination, two themes stand out: first, the debate over the electrical metaphor of field versus flow; second, the notion of stasis versus the notion of enervation. Europe believed for a long time after Maxwell in a world of electrical flows. Ideas of 'action at a distance' or properties of the 'ether' and ideas of an 'electrical fluid' were difficult to unify. The problem of whether or not an 'electrical fluid' was an actual physical entity with the properties of a fluid further compounded the dilemma. Thomson was led to study the whole methodology of physical science, distinguishing 'physical' parts of a theory from 'mathematical' parts. In 1900 he claimed, '[T]here is nothing new to be discovered in physics now. All that remains is more and more precise measurement.'[101] Kelvin's mechanistic approach and refusal to accept Maxwellian field theory reflected the Scottish tradition and impacted on the future of telegraphy. Kelvin, working with Pender, believed in a close association between science, engineering, and industry. This belief led him to distrust mathematical physics and purely academic study. In this aspect, Kelvin and his telegraphy harked back to an instrumentalist Macauleyan vision of the world as *tabula rasa* for continuous social engineering. The flow theory translated in actual practice to needle telegraphy: small surges of current were sent through the telegraph lines. In terms of imagination the agency lay outside of the main body of the system; a brain sending signals. In a paper on vortex motion that Thomson published in 1867, he wrote, 'The mathematical work of the present paper has been performed to illustrate the hypothesis that space is continuously occupied by an incompressible frictionless fluid acted on by no force, and that material phenomena of every kind depend solely on motions created in the fluid.' Across the Atlantic, the system of telegraphy was based on antithetical assumptions, but with almost the same technology. First, the circuit was continuously charged and signals were interruptions to the circuit. Field theory was much easier to accept in this context.[102] Second, while the first vision saw a transforming agency, the second saw it in terms of interruptions to a basic state. This permitted freedom from the obsession with entropy, irreversibility, and conservation of energy that plagued Kelvin, and led him to deny the possibility of Carnot's principle of 'recovery',[103] and thus with his industrial bias became anti-change. Towards the end of his life, Lord Kelvin became as conservative as the cable companies, rejecting Darwin's evolutionary theory, Maxwell's field theory and Rutherford's radioactive theory.

However much the leading experts of the day, Pender, Thomson, and others, tried to deny and arrest change, they inhabited a world in rapid transition. Inhabiting a world still primarily postal in its imagination, telegraphs were imagined in terms of point A connected with point B in linear terms: it was only moments of breakdown that created centres and peripheries in the telegraph system; Fao or Nova Scotia and London were equally important as long as the flow of information kept circulating. In short, the telegraph was a transient system contained within the earlier postal rhetoric and imagination of metropolis and periphery, centres, posts and outposts: it was a mistake that Kelvin made early in his arguments with Clerk Maxwell's electro-magnetic field theory.[104] However, poignantly, 'Kelvin's methodology died with him. It was a methodology that grounded mathematical physics in steam engines, vortex turbines, and telegraph lines; a methodology of industrialisation and Empire ...'[105] In its place the image of the fisherman's net, and of meshes, emerged. Electricity was no longer to be perceived in linear terms but in terms of fields. This revolution in thought and the changes it generated left behind telegraph technology as it grew more divorced from its own roots and became the grand technology. A variety of textbooks on electrical fields and networks began to proliferate by the 1930s. An excellent example would be Ernst Guillemin's *Communication Networks* in two volumes published in 1935. Discussing the two different conceptions, he wrote, 'if we accept this point of view altogether [that is, the seat of the phenomenon of electrical propagation was taken to be solely within the conductor], it becomes altogether impossible to conceive of a flow of electrical energy from one point to another without the aid of an intervening conductor of some sort ... many students are quite wedded to this point of view, so much so, in fact, that to them the propagation of energy without wires becomes a thing altogether apart from other forms of transmission involving an intervening conducting medium. From this angle, the transmission by radio and that by means of a line become not only different in concept but entirely different in analytical treatment and physical interpretation.'[106]

Conclusion

Although the cable industry was immense, it was built over a period of just 30 years. Its importance cannot be underestimated. Eulogies noted, 'Virtually, all the energies of civilised mankind, at once in producing, manufacturing, and commercial enterprise, are regulated, often almost unconsciously, by the cable advices that are ever, without an instant's cessation, darting around the globe on the wings of electricity. ... The wonder of it all is that the whole system is the growth of a single generation.'[107] This was the source of the problem. A huge investment in a particular technology was made without considering that telegraphy might be replaced by other communication technologies. To this vision, telegraphy, like *Pax Britannica*, was something that would last forever.[108] The commitment to submarine telegraphy had been very high for Britain. The 'crisis of imperialism' was not just brought on by an inchoate amalgam of external and internal forces unleashed by late nineteenth century and early twentieth century urbanisation and modernisation and the need of British elites to maintain the status quo,[109] but also, perhaps, was a spin-off from the vast investments made in the telegraph industry and the growing realisation of the need to liquidate these holdings. The imperial crisis was also simultaneously the telegraph crisis. British cable companies resisted switching over to wireless technology on a large scale because of their telegraphic commitments. Through a long period of negotiation the Eastern Cable Companies consolidated themselves into Cable and Wireless Company in 1933.

The other problem lay in the system's need to increase volume of exchange. This allowed the world of information and news to slip beyond any form of control and render the system vulnerable to information panics. In reality, telegraph technology was both much less profitable for the government and much more inaccessible to the general public than was popularly imagined. It served a small elite, who could operate both within it and without it, use it, and popularise the myth of technological success. In the ideological portrayal, access to new technology was unproblematically translated into immediate success and was sold as such to the general masses. Indeed, it could well be argued that the telegraph system, especially the overbuilt and hypersensitive submarine telegraph system that dominated in the 1900s, needed these crises in order to survive and expand. The Australian economic crisis in the 1890s saw the telegraph doing brisk business as capital through telegrams flowed in from the west. In filling vacuums and seeking out crises, the entire system was hypersensitive to the smallest spark. Far from ensuring world peace, what the system assured was a world war. This is not to say that the telegraphs caused the First World War but that it was a symptom of the mounting spillage of the increasing frequency of crises that the conjunction of news and telegraphy had wrought over the turn of the century. As the information order veered out of control and became increasingly prone to panics, so did the technological world that the telegraphs inhabited as it reacted against the beam technology of the wireless and the immediacy of the telephone.

6
Forging a New India in a Telegraph World: Expansion and Consolidation within India

> *The telegraph of India now constitutes a very important link in the telegraphy of the world.*
> Administrative Report of the Indian
> Telegraph Department 1866–7 to 1870–1

The importance of India in the telegraph network

Between 1860 and 1900, the Indian Empire emerged as a crucial strategic element in both the cable and overland telegraph networks of the British Empire. By 1875 India was the main overland link between the West, and the Far East and Australasia. However, less than 70 years later, by 1931, the Indo-European Telegraph Department was desperately trying to liquidate its holdings as it wound up operations. This book examines these 70 years in the context of the imperial telegraph system. By 1874, two main overland lines serviced the connection with India. The first was the Turkish route that went from Constantinople through Diarbekr and Baghdad to Fao. The second was the line run by the Siemens' Indo-European Telegraph Company, from Berlin through Russian Odessa and north of the Black Sea to Tabriz and Tehran. Fao was connected by submarine cable to Bushehr where it met the overland line from Tehran via Isfahan and Shiraz: both these lines were under the management of the Indo-European Telegraph Department of the Government of India. From Bushehr, submarine cables passed via Henjam and Masandam, and Gwadar, further east on the Persian coast, to Karachi. The Eastern Company's submarine cables dropped around the coast of Portugal to touch at Gibraltar, from where they passed via Malta and Alexandria and Cairo through the Suez to Aden, and from there to Bombay. A number of cables run by smaller companies connected Britain across the English Channel to routes passing through Europe and feeding the main telegraph connections to India (please see Map 4.1).

The Indian Telegraph Department controlled the relatively short connection between Bombay and Madras while it ran the lines through the breadth of India from Karachi to Rangoon. The Eastern Cable Company did not yet posses the technological and political expertise to bypass India and Sri Lanka to get to the East, and was therefore vulnerable between Bombay and Madras, from where its near-monopoly stretched to Penang and Australasia. While the Government of India asked for the lease of one of the channel cables and an exclusive line to India or that the government landlines should carry the bulk of the official traffic,[1] the Eastern Company tried over many years to obtain control over this crucial link to the East.

The Eastern Company entered into several price wars with the landlines[2] and opened offices in Liverpool and Manchester, angering the Indo-European Telegraph Department who waited for the allocation of business through the Post Office in London.[3] The Government of India protested several times over the Eastern Companies' discrimination against the Turkish and Persian routes to the East.[4] The Indo-European Telegraph Department was on the point of being rendered useless by an agreement between the Eastern and Indo-European Companies and the Eastern Company's negotiations with the Ottoman Government to control the lines from Constantinople to Fao. The Director of the Department 'raced' to Constantinople to extract a promise that no telegraph concession was to be given to a private company. Finally, a Joint Purse Agreement was reached in 1878 between the Indo-European Telegraph Department, Siemens' Indo-European Telegraph Company, Eastern Group of Cable Companies, the Indian Telegraph Department, and other smaller European parties to divide proportionally the traffic and revenue.[5]

The Indian telegraphs were important to the imperial system and to India's own national trajectory for the following reasons: though it had control over the overland link between Bombay and Madras, it had little say in the distribution of traffic from London to the East.[6] As a result from 1865 up to the mid-1880s the Indian transit rate for through traffic was very high according to the Indo-European Telegraph Department, but the cable companies argued against any reduction because it would reduce their profits. However, the internal rates were higher than transit and overseas rates. While the relative cheapness of Indian rates was emphasised when comparing rates and distances between Europe and

India, the internal telegraphs promised to be self-sufficient while the former were budgeted and subsidised.

It is remarkable that at the point of returning a steady profit, the Indian Telegraph Department was subsumed within the administration as a budgeted department under the Finance Department. The idea that no reduction would be sufficient to offset a corresponding increase of volume dominated policy for over 20 years. As a result, internal rates froze at a comparatively prohibitive rate for the period 1865–85. In contrast, the transit rate, however, continued to reduce during this time. In 1880, the department admitted that the rates charged on internal traffic were 'extremely high in India'.[7] Finally, the inexorable decline in rates affected both international and internal systems, especially after 1885, low rates becoming the norm after the imperial Pacific cable commenced operations in 1902.[8]

This is important in identifying the moments when the opportunities for national economic integration occurred in India. From 1865 until 1885, with the engineering of the massacre of indigenous opium and cotton merchants through the crash of the Bank of Bombay and the ban on opium trade with China, telegraph prices were concentrated on trans-Indian traffic at the cost of, and subsidised by, the rates charged on internal telegrams. Sir Horace Walpole, Under Secretary of State for India, wrote, 'Indian revenues have already borne a large share in the cost of the existing system of telegraphic communication between India and Europe.'

Walpole argued that the inefficiency of the Turkish section robbed the Indian government of the benefit of the Indo-European Telegraph Department while the

> traffic with the Far East, a large share of which is carried through the system thus provided at the cost of Indian revenues, has been developed and maintained at the expense of India, whose merchants are now, by the high tariff enforced on them, contributing more than their due share of the cost of maintaining telegraphic communication between Europe, India, and the Far East.[9]

Indian internal traffic stagnated between 1878 and 1884 because of high rates.[10] After 1888, the decline of rates across telegraph systems in order to survive emerging competition – externally, from the telephone and wireless, and internally, to combat public and private (cable) competition, allowed expansion in indigenous communication, especially in indigenous press telegrams after 1904, leading to increased national cohesion.

The Indian Telegraph Department versus the Indo-European and Mekran Telegraph Departments

The Indian Telegraph Department and the newly formed departments running the two land lines to Europe sometimes entered into contests over matters of finance, precedence and prestige. As early as 1864, Henry Isaac Walton, superintendent of the Mekran Telegraph Department, questioned the authority of Lieutenant Colonel C. Douglas, director general of the Indian Telegraphs, Douglas promptly complained to the government.[11] The government of Bombay sent for an explanation from Walton through his immediate chief, the Commissioner in Sindh. Walton's indiscretion was, perhaps, due to the growing importance of Bombay and Karachi, as they became central points for the large landlines to Europe. Walton was based in Karachi.

Unfortunately, the move proved premature because his appeal was processed through the Indian bureaucracy, which was possibly predisposed towards one of their senior members. Further complicating the issue was the fact that while the minister in Tehran and the political agent in Baghdad rendered their accounts through the accountant general to the Government of India, because these sections of the lines were under the Indian administration, the Resident at Bushehr sent his accounts to the Superintendent of the Mekran Telegraph Department. Given this context, Walton's mistake is unsurprising. Throughout the early 1860s these men had operated in the Middle East and Central Asia as governments within the Government with a wide variety of discretionary powers in matters of diplomatic negotiations, finance, employment, and military action. Walton put in his explanation in writing to the Commissioner in Sindh.[12] Writing from the head quarters of the Indo-European Telegraph Department in Bombay Castle to the Commissioner in Sindh, official opinion noted that 'Walton's representations have failed in explicitness and correctness ... He is warned against the repetition of his remarks with regard to the Director General of [Indian] Telegraphs.' Copies of the letter were sent to Walton to explain his position to the government of India.[13] These were not insignificant exchanges. Walton's application to return to Sindh as the head of the Sindh Telegraph Circle was rejected. As a result his return from Persia to Sindh and India was blocked.[14]

This three-way exchange between the Indo-European Telegraph Department based in Bombay, the Mekran or Persian Telegraph Department based in Karachi and the Indian Telegraph Department centred in Calcutta illustrated the politics behind the growing systems.

On a subsequent occasion, the Mekran Telegraph Department was made more aware of the hierarchy within which it was placed. The Director of the Mekran Department complained that the annual statistical report could not be compiled because the Indian Telegraph authorities had not yet forwarded their reports on the interchange of traffic. He cited the fact that the Indian Director General had forbidden the Superintendent of the Indian Telegraph in Karachi to forward him traffic statistics in contravention of Article XXIX of the Vienna Telegraph Convention.[15] The government replied that all that was required was a succinct report from him for 'incorporation in the Annual Administrative Report of the Government in the Indo-European Telegraph Department.'[16]

In contrast, Goldsmid, the Chief Director of the Indo-European Telegraph Department, complained about the limiting of his travelling allowance while on tour in India, 'as so assimilating my position to the ordinary position of the Director General *in India*.'[17] Goldsmid believed he should get an allowance equivalent to a foreign allowance while in India. In the case of land telegraphs, government control was supreme and attempts by the various departments of the administration to compete which each other were tolerated as long as they showed little desire to expand beyond state control. In 1867 a revision of the Persian Telegraph Department had to be conducted. It was recorded that the sole reason behind the revision was not that salaries were insufficient, the cost of living being low, but that the pay in the Indian Telegraph Department was higher.[18]

The Indo-European Telegraph Department, with support from the Indian administration, made one significant attempt to control the line to Britain and manufacture a monopoly. In 1869 two major landlines were fully operational. The first went from Paris to Constantinople via Vienna and Belgrade to link with the special line from Constantinople to form an unbroken link between Paris and Karachi. The second line went through Italy. The Indian Telegraph Department controlled the line to Tehran. The Siemens Company then ran the northern line through Odessa, Jitomir, and Warsaw to Berlin. The Indo-European Telegraph Department proposed to procure for Indian correspondence three of the channel wires between Britain and the continent. This proposal was made in the context of F. I. Scudamore's (the head of a nationalised and centralised Post Office which also controlled the telegraphs in Britain after 1868) proposal to nationalise the cable lines from Britain and transfer their control from the channel cable companies to government control.[19]

The Government of India suggested that a small office under the Indo-European Telegraph Department be opened in London exclusively for 'India' messages:

> The Indo-European Telegraph Department represents the direct agency which has negotiated the Telegraph Conventions of HM's Government with Turkey and Persia, and introduction of British India into the telegraphic system of continental Europe. If it possessed wires on this side of the channel its administration would obviously be more complete.[20]

It was alleged that the London Telegraph Central Office chose the slowest lines for 'India' messages. Scudamore met Goldsmid and assured him that the Post Office authorities would undertake that Indo-European telegrams would be sent by whatever route may at the time hold out the best prospect of speedy transmission.

The Duke of Argyle at the India Office communicated this to the Governor General in Council.[21] The overall plan of the Indo-European Telegraph Department to secure control of the channel cables through nationalisation by the British government fell through. The anxiety on the part of the British Indian administration to establish a communication monopoly and obtain priority on at least one line for the 'India' messages is understandable, although not justifiable. The Duke of Argyll, using Scudamore's arguments and statistics, pointed out that International and British traffic with India in 1868 was insufficient to occupy even one line.[22] Contradictorily, the demand for a separate and prioritised line from India to Britain remained on the agenda of the British Indian administration even though they had enough proof that the commercial traffic might not be of a sufficient and steady volume to warrant such privilege.

The return of the telegraph manufactories to India: Technological fashions

A crucial spin-off technology began to develop after the 1860s. Articles appeared in the *Journal of the Institute of Electrical Engineers* on the art of packing and transporting instruments to the colony. Metropolitan manufactures, imported in bulk quantities, produced their own sets of problems. Instruments imported from Britain proved useless in India being either unsuitable for use in a tropical climate or they were damaged in transit. As early as 1859, Captain Stewart and O'Shaughnessy asked that Siemens' instruments be examined independently at Berlin and then packed in cases lined with zinc and 'hermetically sealed to prevent damage by seawater.'[23] In 1865, the Indo-European Telegraph Department complained about the 'very much damaged' Siemens' instruments received in Persia 'owing to insufficient and careless packing';[24] Morse instruments fared no better.[25] Though it was 'generally supposed' that England and Europe could supply 'almost on demand any amount [of stores] that can reasonably be acquired at a modest price; but so far from this being the case, there has always been great difficulty and great delay in procuring material of first class quality.' This difficulty, it was argued, was further aggravated by the failure of 'English telegraphists'[26] to realise the 'peculiarities of the tropics.'[27] For example, there was the problem of termites in India endangering certain types of wooden telegraph poles. Before 1857, the Indian lines used a combination of Indian and British manufacture and telegraphic experiment in India, quite independent of European or American experiments, dated back to 1837.

There was a renewal of interest in manufacturing and laboratories for controlling and assessing the quality of instruments imported into India. The immediate issue was the arrival of defective stores from Britain, such as the defective Hamilton's half-standards in the 1860s, and the need to introduce certain modifications for use in India. Telegraph engineers now led the campaign for modification and adaptation. The problem of termites peculiar to India was cited as an obvious example. There were other 'peculiarities' urged as well. The complex of innovation and adaptation across India and Britain cannot be oversimplified into a one-way process of technology transfer from Britain to India. Telegraph technology often drew from its extra-European experiences. Telegraph manufacture and maintenance, transferred wholesale to Britain in the 1860s, returned to India and a succession of European engineer mechanics were appointed to the Telegraph Workshop.

Louis J. Schwendler, previously employed at the Siemens Laboratory in Charlton, was brought out to India after 1866 to take charge of the Telegraph Workshop and Store Yard at Alipur. He was to check the quality of all imported devices and stores; to introduce such modifications as were felt necessary to adapt devices to the Indian lines; a third object was to gradually start a manufacturing process with manufactories or workshops in different parts of India to produce telegraph stores internally. Telegraph technology often drew from its extra-European experiences, and, though accepting Headrick's valuable distinction between 'geographic relocation' and 'cultural diffusion' of a technology, it is important to qualify the first notion and see it as a more complex relation rather than a one-way import-export relationship.[28] Often the personnel for the private cable and telegraph ventures came out of India and the Indo-European Department. For example, Lieutenant Colonel Glover, who represented India at the Vienna International Telegraph Convention, became the managing director of the British Indian Extension and the China Submarine Telegraph Companies.[29]

It must be stressed that this was no longer an experimental phase for telegraph technology. Indian innovations were to be local modifications and given higher status or recognition and in no way challenged the dominance of the metropolitan laboratories. Moreover, the technology had been tamed, homogenised and standardised in the technopoles. What subsequently took place was classified as local innovation and modification even where it was thought important enough to adopt universally. Schwendler and Muirhead perfected the Duplex system for the Indian lines, which could be worked in both directions simultaneously. This was first introduced between Calcutta and Bombay in June 1874 and subsequently between Madras and Bombay in February 1875.[30] The annual administrative report of the department for 1870–1 recorded that, 'Satisfactory results have been obtained by ... enlisting in its ranks men of sufficiently high education to understand their application [of the latest scientific discoveries of Europe] and with originality to modify and improve them to suit the peculiar circumstances in this country.' It noted that for some time, all stores that were required to be electrically perfect were put through severe tests at the Telegraph Workshop and Store Yard, in a special testing room equipped with the latest instruments for that purpose. It concluded, '[B]y its adoption not only is the issue of perfect stores guaranteed, but being manufacturers ourselves we obtain a complete control over the work done.'[31] It was also noted that the capacity of the Telegraph Workshop had been increased with the 'erection of an excellent wire-twisting and cable-making machine.'[32]

This return of the laboratory to India signalled its importance in the total system. Echoing O'Shaughnessy in the 1850s, the director general stated, 'For India we require a much lighter apparatus than we use in England and must avail ourselves of bamboos and other natural produces of the country. We cannot be wholly independent of Europe, but we should get as little as possible from her, and that of the commonest description, that is to say, we should look to India for everything but common iron wire.'[33] However, it must be emphasised that this was not a simple return to the early days of telegraphy in India.

Telegraph signallers and engineers were now trained and recruited mainly in Britain and every year four of the 'most intelligent' officers of the Telegraph Department were sent to Europe for further training, which included a regular course in modern physics.[34] The same report was to paradoxically conclude that the 'large, strong, electro-magnetic instrument introduced by Siemens is the best adapted for India.'[35] These internal moves also sparked off rivalries with private cable and manufacturing companies and other telegraph departments.

Trajectories of surveillance

The telegraph appeared to promise several services to the British government in India, and the Government of India adopted the principle to control all telegraph lines after 1874.[36] Perhaps, the British government imagined that the telegraph, being a state owned communication system based on technology, would transmit messages that were rational and controllable.[37] This contrasted with the earlier reliance on indigenous informants and their networks, which were perceived as irrational, spontaneous, susceptible to rumour, and highly contagious and volatile networks of human communication. The telegraph, replacing indigenous informants after 1860, was to be the crux of a rational and exclusive system of information circulation. By the 1870s the government also thought that the telegraph would provide a means of accessing indigenous communication and to keep watch over unrest, which many senior civil servants such as A. O. Hume, were convinced lay simmering beneath the apparent placidity of *Pax Britannica*.

The contradiction faced by the colonial state between suppressing seditious propaganda and the need to know of its existence was reflected in the recurring measures adopted simultaneously to restrain and promote the press (a spin-off industry of the telegraph after 1860), in particular, the vernacular press. The Duke of Buckingham argued for the usefulness of the vernacular press in indicating the 'under-currents, which may be running through the mass of Indian population'.[38] Surveillance through the means of the telegraph and physically through telegraph offices and personnel also became an important necessity for the increasingly electronic empire. Shortly after India and Britain were joined by telegraph, along with the disciplining of telegraph personnel, a series of measures were adopted to control, contain, and survey communications.

Telegraph reforms coincided with the reform of the police in India. The reform of the administration after 1858 laid down the basic structure of attitudes and policies that governed much of the second half of the nineteenth century. Along with the penetration of communication an expanded police force was needed to control, protect and observe more people. The Indian Police Service might be said to date from the Police Act of 1861; it was the beginning of a civil police as distinct from the earlier military police.[39] Reflecting the trend in the telegraph establishment, and almost all aspects of the administration, all officers above and including the rank of superintendent of police, were mostly British with rare exceptions until 1919. However, the contradiction between employing British personnel and the need to access the undercurrents of Indian unrest and authentic indigenous knowledge remained a fundamental problem for the British government well into the First World War. Similarly, the need to maintain surveillance over indigenous communication complemented the government's need to control its own information incontinence; often the same law operated both ways.

In 1873 the central government passed strictures on local governments for giving a general sanction to examine telegraph messages under Section II of Act VIII of 1860 allowing the divulging of telegraph messages. The message in question had to be specified in every instance by the secretary to the government or a similar Indian Civil Service authority. It pointed out that the

> Act above referred to is not addressed to the purpose of giving the Government the power of examining telegrams, but to that of imposing severe penalties on the Officers of the Telegraph Department if they divulge a message. A sufficient order may relieve them from the penalty but the law does not bind them to obey any such order ... [the telegraph officer is] entitled to be placed beyond doubt that the order applies to the message divulged.[40]

Obviously, the local governments had gone beyond the original intent of the law and needed to be reminded of its original premise. Similarly, the central government pointed out the 'unnecessary' use of the term 'confidential' because such a message was dealt with personally by the head of office, 'who allows no one else to see it, and sends all the signallers out of the room, while he himself signals it towards its destination.' This led to delays as all other business on every wire was temporarily suspended until the message was transmitted.[41]

Imperial anxiety about correct information was reflected in a two-pronged strategy: first, to provide British officials with an untainted and authentic archive of information; second, to provide an equally chimerical authentic news service to the media, which would then act as a propaganda vehicle for the government. Accordingly, an office was established in 1864 to provide 'loyal and important' newspapers with government briefings, access to select records and documents, and official news. Annual and monthly digests of abstracts and reviews of vernacular publications by the intelligence services of the local governments, collated and synthesised on a central level after 1870 by the Special Branch of the Thagi and Dakaiti Branch, were circulated among the bureaucracy in order to provide means to access vernacular and indigenous public opinion. The surveillance offensive was reflected, for example, in the Act xxv of 1867 'for the regulation of printing presses and newspapers, for the preservation of copies of books printed in British India, and the registration of such books', which provided for an overall surveillance of printed material. Similarly, Section 26 of the Indian Post Office Act allowed the government to intercept postal articles and Section 19 of 1878 of the Sea Customs Act authorised interception of articles of import ranging from pamphlets, books, 'French novels' and pornography, to guns and ammunition.

Recurring issues of censorship and translation

Censorship of the Press and its many features have been much discussed; administrative histories, such as by Barrier,[42] have been complemented by analyses questioning categories of proscription, investigation of figures of sales and usage in libraries, distribution of presses, and print runs.[43] However, much of this genre has ignored the physical means of control over the transmission and dissemination of information; for example, the cost and the kind of paper, the locality of its manufacture, and changes in post and telegraph rates, in terms of weight for the former and number of words for the latter.

The 1818 Deportation Regulation, resurrected in 1908, originally targeted the Anglo-Indian press clamouring for the repeal of the law prohibiting non-official purchase and settlement of land. James Silk Buckingham, the then most dangerous spokesman and publicist for a Creole and settler movement, was neutralised before his deportation by the manipulation of postal rates, charged on different categories of the weight of paper. It needs to be remembered that thin, handmade, paper, manufactured indigenously, was charged double rates by the government post office in India even though this paper weighed less, was available locally, and was cheaper to buy in the market. The amount charged by the government post office for sending a letter on the thicker, imported, and heavier European paper was half the rate charged on Indian paper. This unfair and prohibitive charge lasted until as late as 1837. Subsequently, the uniformity of rates provided equal access to the coercive postal monopoly of the British government and the introduction of paper mills after the European style around the middle of the nineteenth century heralded the doom of indigenous paper manufacture.[44]

Although dialogue in the public sphere between the rulers and their subjects, such as astronomy, are rightly celebrated,[45] it must be remembered that this was also a highly controlled and fragile economy of information. After the assassination of the Earl of Minto in the Andamans by, allegedly, a Wahabi fanatic, Sher Ali, the British government examined with urgency the networks that it had lost sight of, and, ironically, facilitated. Wahabi networks, stretching from Turkey and Persia to Patna and the Andaman, were again the focus of anxiety and surveillance after letters from Wahabis in Patna were found addressed to the convict.[46] Officials recommended that the Thagi and Dakaiti Department needed fresh strategies and enhanced funding to control criminals using the railways and the telegraphs. The Vernacular Press Act of 1878, promulgated by Lord Lytton after Minto's assassination, was also formulated under the urgency of the looming war between Russia and Britain over Turkey and Persia, was intended to be a measure to control the vernacular press.

The law demanded unspecified securities from proprietors of newspapers and several of the most important vernacular journals folded operations. However, like all such acts it achieved little that was practical. In effect, it antagonised the Liberals in Britain by 'gagging' the press and in India they faced opposition from educated indigenous public opinion. This liberal combine was supplemented by more conservative bureaucratic opinion, which held, like the Duke of Buckingham, that it was more important to know, ventilate, and reveal, than to suppress Indian public opinion. Anglo-Indian newspapers commented on the fact that 'a prosecution invariably brought fame and fortune' to the editor and 'helped to circulate far and wide the opinions which must otherwise have died in their native obscurity. Can anyone doubt', it was asked, 'that the influence and importance of the class of newspapers against which the Bill is directed, will not be considerably enhanced by the notoriety of official notice or censure?'[47] Similarly, the *Daily News* warned that discontent might disappear from newsprint but would become 'all the more virulent and intense when it is disseminated under the seal of secrecy through the Post Office'.[48]

Indian public opinion used the critical Anglo-Indian press to argue that the Vernacular Press Act, so far from preventing 'disaffection', would actually be 'directly provocative of dangerous sedition'. Vernacular newspapers were distributed by post so the 'native publisher has only to lithograph his seditious articles on thin paper and send them to his subscribers in envelopes' and the government would be 'entirely powerless':

> It [the Government] cannot break open letters, or at least cannot act on such a violation of public confidence ... it is notorious that this was done by the mullahs and others before the Mutiny, and that the Persian newsletters were far more dangerous than the worst of newspapers have ever been. The effect of compelling publishers to circulate papers in secret is to liberate editors from all responsibility to the law ... The disease would therefore be driven in and the natural development of the Press arrested, in order to make it, or enable it to become, with impunity, twice as objectionable as before. Is it worth it to create this precedent ... in order to compel a few native publishers to print their seditious articles upon thinner paper than they would otherwise use?[49]

Simultaneously, rumours began to proliferate and comparisons with the impending signs of 1857 and similar events were drawn. Locusts were reported to have appeared in Mysore as a 'bad omen to the prospects of the country, as they once made their appearance just before Coorg and Seringapatam were taken by the British.'[50] They were soon reported to have appeared on the borders of Bengal, in Tirhut and Malda districts, after having ravaged large parts of the Madras Presidency.[51] The function and symptom of the plague, described by Artaud[52] and others, are relocated not to the individual but ascribed to the body politic, both local and central: the larger Indo-Asian body politic was always quick to exhibit and read symptoms of malaise. A Eurasian or Portuguese person fired a shot into the carriage of Lord and Lady Lytton in Calcutta[53] and Lytton complained of the animosity of the Anglo-Indians 'reflected in private gossip and newspaper articles about Indian Affairs in London.'[54] The problem of defining sedition recurred throughout British rule. Surendranath Bannerjea, in his speech to at the town hall meeting called by the Indian Association on the 17 April 1878, accused the government of 'mistranslations' and of reading passages 'out of context', and alleged that the government's charges of 'sedition' and 'disaffection' were 'chimerical and had no foundation in fact'.[55] The Maharaja

Holkar, eager to demonstrate his loyalty to the British government, imprisoned the editor of the *Malwa Akhbar* for criticising the Act.[56] When the storm had passed and the Act was all but dead in reality, Holkar accepted the editor's explanation that the allegedly seditious passages in his newspaper were translations from the Anglo-Indian newspapers and Reuters despatches.[57]

Widespread public meetings and memorials protesting against the Vernacular Press Act forced the government to adopt several defensive measures. The Government of India passed new rules regulating the transmission of petitions to the empress and secretary of state. Petitions had to be forwarded through local governments, which reserved the right to withhold petitions. The Government of India passed strictures on the conduct of the governor of Bengal, Eden, for his 'officious zeal'. The Government of India backed down and promised that the Act would cease to be applied though it would remain on paper. It offered the olive branch to the vernacular press by extending to them press rates for telegrams previously afforded to the Anglo-Indian and English-language press. They could now subscribe to Reuters telegrams and a vernacular press commissioner was appointed to supply vernacular newspapers with 'authentic information'. Not that Reuters was considered to be invariably reliable by the government, because Lytton wondered how much the Baron Reuter received to circulate the information that the Emperor of Russia had recognised the assumption of the title of the Empress of India in 1877 by Queen Victoria.[58]

It was noted in the press, in the resurfacing of a bill modelled after 34 and 35 Vict., Cap. 112 in British law for the prevention of crime in the Irish context first proposed in 1873, that the development of railways and the increased means of procuring employment at a distance from their homes had rendered 'loose characters' more 'able and willing' to leave their villages and move elsewhere.[59] Arguing for the establishment of a 'system of surveillance', Sir James Lyall, Commissioner, and C. L. Tupper, Secretary, the government of the Punjab, stated that the rules 'should involve a minimum of Police interference consistent with a real supervision over movements and mode of life of persons subjected to surveillance.'[60] The Government of North Western Provinces and Awadh, recorded

> [t]he improvements in communication, effected chiefly through by the railways and the telegraph, have greatly facilitated the operations of those who are in the habit of leaving their homes to commit crimes at a distance and returning with their plunder. It was

accordingly thought necessary to strengthen the hands of the police, by giving to those charged with prevention and detection of crime some compensation for the facilities now afforded to the more adventurous criminals by the improvement of communications.[61]

The expansion of communication systems, which often anticipated indigenous mobilisation and unrest, was a source of anxiety for the British Government in India. Elaborate systems of surveillance simultaneously promoted, created, and re-invented networks of disaffection that, although already in existence, were substantially transformed by the new communication technologies. As communication networks expanded the anxieties of the British government in India increased.

Imperial 'high noon': Curzon and India 1880–1905

The Lords Curzon, Cromer, and Milner represented a new imperialism in the colonies after 1880: they were the satraps or proconsuls of the British Empire. Cromer ruled Egypt, Milner South Africa, while Curzon asserted imperial and vice regal paramountcy in India. Ironically, these men were advocates of 'old style'[62] imperialism, grandiosely echoing Roman terminology. The triad were the antithesis of Liberal 'New imperialism'. They were conservative, despotic, and dedicated to the notion of imperium. Their authoritarian regimes, in contrast with the emerging Labour ethos in Britain, highlighted the importance of the telegraphic hubs of the British Empire in the late nineteenth century. It was also a question of *imperium in imperio* (state within a state): the India Office was the only department of the British government with an army and an independent budget. The India Office was an independent player in imperial politics, free from Parliamentary scrutiny until John Morley became Secretary of State in the 1900s.

Control over the Suez Canal and southern Persia, political stability for telegraph cables girdling Africa and the original project to telegraphically connect the Cape to Egypt over land, the importance of India as a means to reach Australasia and China, were the reasons for the iron control and expansion exerted over and from these areas. The Boer War and the occupation of Egypt reflected this need to exert sovereignty over areas and routes perceived to be vital to an empire heavily dependent on the telegraph system. Lord Roberts believed it to be crucial to push through with road, rail and telegraph, to ensure British control over the Afghan frontier.[63] The cheapest and most practical of these was the telegraph.

Lord Curzon saw British India as the 'true fulcrum of Asian dominion' and realised that Britain's international hegemony was based on control over India. The removal of a British presence would not only endanger British dominion but would also, according to the Conservative belief, lead to the internal dissolution of India. Thus, the two fortunes were inextricably tied together: India as an entity would never survive without Britain, and Imperial Britain could not survive without India.[64] Curzon declared, 'As long as we rule India, we are the greatest power in the world. If we lose it we shall drop straightway to a third rate power ... Your ports and coaling stations, your fortresses and your dockyards, your Crown colonies and protectorates will go too ... an Empire that has vanished.'[65]

The obvious and apparent reason was the military might afforded by India. By the end of the nineteenth century, 40 per cent of Indian revenue went to the military.[66] The Indian Army, with its huge reserves, enabled Britain to be the foremost military power in the world. The Indian Army was deployed on numerous sites outside of India. For example, it was used in China in 1839, 1856, and 1859; in Abyssinia in 1867; in Egypt in 1882; in Nyasaland in 1893; in Sudan from 1896 to 1898: the examples can be multiplied. Extra *batta* was granted in 1868 to officers and men of the Indian Telegraph Department who had served during the Abyssinian campaign of 1867[67] and Lieutenant St. John reported to the Government of India on the material and operations of the department during the war.[68] In the 1898 scramble for Chinese ports, Russia occupied Port Arthur, Britain Wei-hai-wei, and Germany Kiaochow. Writing of the relief of Peking during the Boxer uprisings of 1900, one of the members of the besieged legations, B. L. Putnam Weale, recorded his immense relief when, at the point of exhaustion he recognised the

> ... smell of India! ... hundreds of native troops were filing and piling arms. They were Rajputs, all talking together, ... and demanding immediately *pane, pane pane* all the time in a monotonous chorus .... I ran on, once more choking a little, and with a curious desire to weep or shout or make uncouth noises ... I did not care; the relief had come.[69]

In rapid response to the Boxer uprising, the Eastern Company laid a cable from Shanghai to Chefoo and Taku, the ports closest to Tientsin and Peking, and Wei-hai-wei, the main British naval base, and generally extended its sphere of operations to northern China.[70] The company rapidly rebuilt the overland telegraph destroyed by the Boxers, the 'antipathy to being dependent on the Chinese in so important a matter as telegraphic communication was then particularly strong ...'[71]

Curzon, on his part, sent nearly 20,000 Indian troops to relieve the besieged Legations in Peking. The 13,000 British Indian troops sent a year earlier to South Africa were equally important in preventing the fall of Natal in 1899. Lord Milner recorded that had it not been for Curzon, Boer flags would have flown over Maritzburg and Durban. It is doubtful, whether Britain could have recovered her position in South Africa had she lost Natal.[72] The Eastern Telegraph laid a cable to South Africa via Ascension Island and St Helena.[73] Telegraphic connection not only served the strategic and mobilisational needs of the empire but also provided rapid news about events on its outskirts. The telegraph followed the British flag as much as the flag followed the telegraph, that is, the telegraph needed secure land and sea routes so governments were drawn into provid ing military support. The British military used a system of telegraph code addresses during the Boer war and in Egypt that bordered on the extreme: Major General (then Colonel) Thorneycroft was unhappily '*Unbuttoned*', the Chaplain General was predictably '*Meekness*', while General Clery, when Chief of Staff to Wolseley in Egypt, discovered that his label was '*Castration*'. When Clery's objections were conveyed to the Duke of Cambridge, then Commander-in-Chief, the Duke is said to have replied, 'Quite right too, so would I'.[74] In short, the telegraph was a continuous presence in every imperial campaign after 1850.

Many in Britain and India feared British Indian pan-imperialism under Curzon. Writing after the War, Montagu commented:

> And then there is the rounded Lord Curzon, who for historical reasons of which he alone is the master, geographical considerations which he has peculiarly studied, finds, reluctantly, much against his will, with very grave doubts, that it would be dangerous if any country in the world was left to itself ...[75]

Lieutenant Colonel Francis Younghusband echoed these sentiments when he wrote to Curzon 'Tibetans are not a people fit to be left to themselves between two Great Empires. They have to look to the one or the other, to us or the Russians, for protection. We cannot afford to let the Russian influence prevail ...'[76] British Indian involvement, financially, militarily, and politically in far flung corners of the empire was not solely because of Curzon's megalomania but a symptom of the new international order that emerged because of the telegraph annihilating time and space so that no corner of the globe seemed inaccessible and because of India's pivotal place within this system. Curzon declared his objective as 'Anglo-Indian Imperialism. I want the natives to see the benefits and share in the glory.'[77]

Even before he left for India, Curzon was widely regarded as a champion of an aggressive policy in west Asia, which could provoke war with Russia. He divided the province of Punjab to create the North-West Frontier Province as a first step to exerting control over the frontier.[78] Advocating a clear policy that sent appropriate signals to Russia regarding its designs on the Persian Gulf, he wanted a statement of British interests in the region, which, from the mid-nineteenth century, dominated the region not so much because of trade and the reduction of piracy,[79] but because of the telegraph along southern Persia. As a result of Curzon's continuous despatches the British government informed the Persian government that while she supported the integrity and independence of Persia, Britain 'held such interests in the south of the country that she could not accept the rivalry of another power there.'[80] These interests were primarily telegraphic and by 1903 Britain proclaimed a policy similar to the Monroe Doctrine in the Gulf whereby any encroachment by another power in that area would be seen as an act of hostility. Curzon, who was 'delighted' by this bold statement of a policy he had advocated for, embarked on a magnificent 'flag-waving' tour of the Persian Gulf now reduced to the position of an British Indian lake; St John Brodrick, Secretary of State for India 1903–5, mockingly rechristened the Persian Gulf as the 'Curzon Lake'. Of course, as Lord Curzon sailed 'majestically' up the Persian Gulf in a curiously ambivalent assertion of both British and British Indian imperial dominance over the area, he was accompanied by two telegraph and despatch boats to report and record his triumphal progress.[81] He similarly supported an intervention in Tibet, again incurring the ire of the Home authorities, especially St John Brodrick, the Secretary of State for India.[82]

Lord Curzon extended the frontiers of British India as much as he exerted British and Viceregal supremacy within India, and admitted to being a 'tyrant' by 1900.[83] Ironically, Curzon believed that the spread of telegraphy had proved detrimental to the quality of the Indian Civil Service and to the knowledge, expertise, and initiative of the man-on-the-spot. The Civil Service Code and the telegraph left the individual administrator a puppet in the hands of the local government, while steam, post, and telegraph brought Britain much closer to him. Thus, the man-on-the spot saw his time in India as a period of exile and looked to Britain for news and entertainment at the cost of genuine involvement in his work in India.[84] His zeal for reforming an overburdened and tortuous bureaucracy went to extremes: he thought there was only one secretary of a government department who could write decent English, and he imported several 'experts from Britain to examine and report, and in some cases direct, his various projects.'[85] Curzon's assertion of authoritarian might over Indian policy and his championing of the interests of India against an encroaching India Office and the larger Empire was contradicted by his regular use of metropolitan expertise on committees set up to reform branches of the administration. Obviously, the entrenched British bureaucracy in India resented the incursion of 'outsiders' from Britain who worked under the halo of universal expertise. The battle between a universal notion of control and efficiency, rational and scientific, and the intuitive insights of the man-on-the-spot familiar with a unique India, came to a head under Curzon's reign. An indefatigable worker, Curzon, that very 'superior person', perhaps, took himself entirely too seriously, and the combined attack of the Anglo-Indian press and the offended bureaucracy during the early years of his reign spiralled into political extremism within the Indian National Congress and revolutionary extremism outside of it.

The Curzon-Kitchener controversy over leaking of confidential state information

Lord Curzon reformed several key areas of the administration: British Indian frontier policy; the higher education system; police; the creation of a new department called Commerce and Industry that subsumed the telegraphs; finance; territories under administration; the military, 'reforming the machine itself'.[86] For most of these reforms, in specific areas and enterprises as well as for holistic reform he imported experts. Lord Kitchener, the 'hero' of Sudan and South Africa, was one of his prize imports. Kitchener had, for some unknown reason, acquired the status of a myth in the popular and influential imagination of Britain, what the *Pioneer* would later describe as the 'fetish worship of Lord Kitchener',[87] and Curzon wanted him to undertake military reforms in India as well as add to the Viceroy's political prominence in British political circles as the Imperial Proconsul controlling the 'greatest soldier in the Empire'.[88] Curzon fully believed that Kitchener would become a part of the Viceregal and official 'family' and expected to form a 'dominant-dependent' relationship with him.[89] Lord Kitchener wanted to remove the Military Member from the Viceregal Council, arguing that it represented dual authority. The diplomatic manoeuvring in Britain by these adroit players of British Imperial politics ended in Curzon's resignation over an issue that was, perhaps, little understood by the general public. Much has been written on this controversy from a variety of angles including military efficiency[90] and the 'personality approach',[91] but this section focuses on the issue of what constituted official and legitimate channels of communication and information.

Both Curzon and Kitchener had independent access to the Secretary of State and the cabinet, the matter of the military member of the viceroy's council not being Kitchener himself or appointed by him became a question of hierarchy and precedence. Kitchener, surrounded by a coterie of military assistants including Birdwood,[92] was reluctant to play second fiddle to Curzon in India. Having earned his reputation in the post-1870 imperial expansion in the margins of empire, he saw India within the larger imperial context. Throughout the period with Curzon he maintained independent communications with select officials in the India Office and other political institutions in Britain including the cabinet. It was a question of the most convincing representation of either case before the Parliament and the public in Britain.[93] At his farewell dinner at the Byculla Club in Bombay, Curzon stressed the two principles over which he had resigned: first, the precedence of civil over military administration; second, 'the payment of due and becoming regard to Indian authority in determining India's needs.'[94] Thus, as Cohen shows, the two differed in their understanding of imperial necessity and India's role within the Empire: Curzon believed the future lay in dual leadership of Britain and India, while Kitchener thought the Empire and Britain, the Committee of Imperial Defence formed by Arthur Balfour in the 1890s in particular, should be the controlling instrument.[95] John Arthur Godley, Permanent Under Secretary of the State for India (1883–1909) wrote to Curzon that 'the *real* Government of India is in the House of Commons'.[96] Kitchener viewed the British Empire as first British and of Britain therefore to be ruled by it rather than an empire serving to secure India and routes to it.[97] Curzon disagreed and saw executive authority as shared between the Secretary of State and Cabinet at one end and the Viceroy and Council at the other. Stressing the need for the supremacy and trust of the man-on-the-spot, the man famed for ignoring popular opinion argued that the Viceroy had to deal with the growing Public Opinion in India and that it was 'impossible to attempt to rule India' from England through the telegraph.[98]

The dispute over military administration was conducted through private correspondence, leaks to the press, and administrative Blue Books. Thus, what was purely a matter of administrative efficiency became a public confrontation between Curzon and Kitchener, and the press in India and Britain became important players in this politicised dispute. Lord Kitchener circulated his note on the reform of the Military Department to the Committee of Imperial Defence. Curzon protested to Balfour that such proposals could not bypass the Secretary of State and the Government of India, and that it was, in any case, a matter of internal administration and not something that the committee had power over. Colonel Herbert Mullaly, a member of Kitchener's staff, was the source of the leak. Regular telegrams in private code were exchanged between Kitchener and the private office of the Secretary of State. However, Kitchener felt nervous and changed to the Indian Army 'Y' cipher for greater security in 1905. In turn, Curzon was offered copies of Kitchener's coded telegrams to the War Office by a 'native gentleman of good position'.[99] Lady Salisbury and other influential individuals close to the cabinet and the king were enlisted in Britain to promote Kitchener's cause, and by 1905, H. A. Gwynne, editor of the *Standard*, and Colonel A'Court Repington, military correspondent of the *Times*, were also won over. Thus, channels of private, official and press communication were entangled in the Kitchener-Curzon dispute.[100] To exert control over official information leaks, Curzon passed the Official Secrets Act of 1904. Widely regarded in the vernacular and Anglo-Indian press as a measure to control them, the Lieutenant Governor of Bengal expressed its true purpose:

> I have found papers going to the Press marked confidential, I have notes relating solely to the conduct of officers finding their way to the Press, I have found demi-official letters which I myself have written finding their way to the Press ... this is due very largely to the fact that there is *no conscience in regard to the communication of information.*[101]

Curzon lost the battle and resigned over an issue few in the general public could understand in 1905. His reign in India was regarded as both the 'apogee of the British Empire and its decline'.[102]

In this context, it would be futile to argue for metropolis and periphery or British and Indian since the drama unfolded simultaneously in both countries: Indian issues caused major political conflict in Britain, while British political swings impacted on India. Lord Kitchener and his coterie managed to get away with their tactics with Lord Curzon, but Curzon's successor, Lord Hardinge, caught them out. Commenting on the leakage of the proposals to reduce military expenditure in India, which had reached the *Standard* and the *Times*, Hardinge wrote to the

Secretary of State: 'I have now traced it to the Chief of the General Staff, General William Haig, who within the last two or three weeks, has committed a very serious and flagrant breach of trust.'[103] He subsequently informed the Secretary of State that he had additional confirmation that 'information from here was being transmitted privately to the War Office in London and from there it got into the Press.'[104] This welter of growing rapidity of information circulation and increasing information incontinence of the state was a function of the telegraph network annihilating time and space; what Lovat Fraser described as the 'subterranean channels of international intelligence.'[105]

The telegraph and the standardisation of time

Recent research has shown that telegraphy's 'commercial success demonstrated that the economic value of a message depended not only on its content, point of origination, and point of destination, but also on the expected mean and variance of transmission time'.[106] In P. G. Wodehouse novels, telegrams play a central role, symbolising the speed of modern communications and the consequent rapidity of alarums and incursions. For example, in a typical exchange between Bertie Wooster and Jeeves:

BERTIE: When you have brought the tea you had better go out and send him a telegram, telling him to come up by the next train.
JEEVES: I have already done so, sir. I took the liberty of writing the message and dispatching it by the lift attendant.
BERTIE: By Jove, you think of everything Jeeves!
JEEVES: Thank you, sir. A little buttered toast with the tea?
ROCKY [the recipient of the telegram]: [W]hen your telegram arrived I was just lying down for a quiet pipe, with a sense of absolute peace stealing over me. I had to get dressed and sprint two miles to catch the train.107

The telegraph impacted upon and changed lives in tangible and intangible but fundamental ways. The most basic impact was on how people thought of time and space, since the telegraph claimed to 'annihilate time and space'. The telegraph, much more directly than the railways or steam, necessitated a global coherence of time. Business being transacted over vast distances, the telegraph demanded a centralised and standardised time instead of the prevailing freedom of local times. Although much has been written on what constitutes time and its meaning, the historical processes of the construction of time have been relatively neglected. Indeed, perhaps, the sophisticated hermeneutic dialogue over time and presence presupposes a standardised construction of time.[108]

Revealingly, very early texts dealing with the telegraph lapsed into long discussions of western versus eastern methods of reckoning time. Kalidas Maitra in the first book in Bengali on telegraphy gave elaborate conversions between *Puranic* or classical Indian time reckoning and western time. Arguing for the superiority of the western division of the day into 24 hours over the 60 Indian divisions or *dandas*, he more subtly argued that the *Agnipuran* was the source of the division of the day into *bela* or *hora*: the origin of the word 'hour'. Maitra provided a guide to both western and Indian time calculation as well as homogenising time calculation. He was conveniently presenting two major time-reckoning methods at the expense of the many others that existed. He also promised a separate book on the origin and calculation of time, especially for astrological purposes.[109] Switching to Madras standard time in 1861 started the process of homogenising and standardising time in India internally.

Not long after O'Shaughnessy's departure from India, the Department complained strongly against the 'uncertainty' of 'civil time':

[The] considerable difficulty for some time when tracing delays owing to the uncertainty at all time attending the mode used in civil life of expressing time ... an inherent uncertainty in the mode of expressing civil time, which is elsewhere remedied by supplementary expressions, but which the necessity of brevity excludes from use in the Telegraph.[110]

The need to calculate delays and compute the efficiency of the Indian Telegraph Department in statistical comparison with other telegraph departments and companies combined with the peculiar problem of signalling midnight and noon. The public and the signaller's frequently confused 12 noon and 12 midnight and even Bradshaw's railway guide used 12 noon, the phrase, to signify day instead of night. The Indian Telegraph Department proposed to switch to a continuous timescale, that is, it would reckon time not in terms of divisions of 12 but in terms of 24. So 5 p.m. would now be 17 hours; a convention quite unremarkable today though Douglas had to cite articles in the western press[111] to claim legitimacy for this method and sought permission from the government to make this change in telegraph time reckoning. Little did he realise how changes in telegraphic time reckoning meant a reordering of time itself.

The construction of a general and arbitrary telegraph time occurred the next year. The annual report noted that because of the 'Departmental inconveniences' resulting from the 'Civil mode of time reckoning', along with the change in time reckoning which was introduced to avoid them, there was a further change: 'the introduction of an arbitrary time, as the Telegraph Time for all India.' Madras time was chosen for several reasons. First, the Madras longitude was nearly equidistant from Calcutta and Bombay. Second, the adoption of its time involved the minimum difference between local and telegraph time. Third, as Madras had a government observatory, '*true* time from it can more readily be obtained than from any other station in India.' Fourth, madras had the additional advantage of being a seat of government and the headquarters of a departmental division and circle.[112] There could also have been other factors at play behind the decision. In the rivalries between the Presidencies towns, Madras appeared a neutral option. Moreover, much of the scientific and laboratory work was conducted in the south during this period, possibly because of its relative stability in comparison with northern India and the recent experience of the mutiny, and its proximity to the equator.

The increasing unification of the first electronic network had a more immediate and fundamental urgency.[113] The structure of time was fundamentally changed. The Telegraph Convention in Rome adopted the zero to 24-hour approach.[114] This happened over a period of time and reflected the increasing homogenisation and proliferation of the telegraph system. At midnight on the 1 July 1905 all telegraph clocks in India were synchronised: Standard Time was introduced in India and Burma.[115] Its origins lay in a note in the Meteorological Branch under the Department of Revenue and Agriculture submitted the year before.[116] In reports from the local governments several notes of caution were sounded. MacLagan noted '[T]he apparent readiness of Calcutta but not of Bombay and Karachi to adopt the five and a half hour's standard [from Greenwich Mean Time] for local time ... it appears that if the Standard Time is adopted in Bombay, Calcutta and Karachi, it will for all practical purposes be universal in India.' He suggested a vigorous campaign by the government through the introduction of Standard Time in post offices, schools, government offices, and the firing of midday guns.[117] Denzil Ibbetson disagreed in his note and warned that any 'appearance of trying to force the hand of the non-official community will do more harm than good.' He pointed out that it was the Lieutenant

Governor Sir John Fergusson's attempt to 'hustle' the Bombay people by introducing Madras Time in all government institutions that led to the failure of the experiment and most of the private sector and the banks had kept to Local Time.[118] Karachi managed to drum up public support from its local elites[119] but the Bombay business community presented a different case. The government of Bombay warned that although the 'bulk of opinion' supported the replacement of Madras Time with Standard Time on railways and telegraphs, in the city of Bombay and in Karachi there was 'a strong preference for Local Time.'[120]

Protests started out among workers especially telegraph workers and postal deliverers who were the first to be hit along with the railway workers. However, railways were comparatively less affected with an approximately eight minute adjustment from their previous Madras Time schedule. The *Bombay Samachar* wrote in an editorial entitled 'The unpopularity of the Standard Time with the masses' that the 'Government will earn the blessings of the native population by restoring to the City its old time.'[121] The delivery establishment in Karachi and Sindh went on strike and complained in the press that the new railway schedules had forced them to do three deliveries a day when they had previously done two. 'Owing to the change in the Railway Time-table ... Posts have to be delivered thrice instead of twice, yet there is no increase in the staff.'[122] Sporadic strikes began to break out in the public sector and in the mills. The panic issuing after 1900 was as much a systemic crisis as disparate information panics, nationalisms and workers movements constituted it because they operated within the same telegraph system.

Was it an information panic or systemic crisis: Conclusion

The term 'information panic'[123] maps how peripheral conflicts and panics allowed the colonial state to interfere in and extend the margins of rule. I am extending the use of the term to include its systemic rather than erratic aspect. I argue that panics were an integral part of the system and that panic on the margins or at the centre reflected a holistic and systemic panic. Apparently discrete, peripheral events appeared in the international media taking on a snowball effect, that is, once started events and panics achieved a velocity of their own. For example, in 1907 the Government of India put the provinces of Bihar, Bengal, Uttar Pradesh and Central Provinces on extreme alert. Mysterious markings appeared on the trees in many of these regions, and these were read as a signal for unrest. Roving bands of sadhus and sanyasis were suspected, and in some cases, arrested. The spectre of the mysterious *chapati* of 1857 spreading the message of revolution in the countryside was resurrected. Sanyasis and Sadhus were a constant source of anxiety for the colonial state and on this occasion hundreds of them were arrested. An editorial warned that '... if every Sanyasi sees that he is suspected of being a conspirator he may be forced to be so in grim earnest.'[124] This proved to be a rumour and the expected upsurge in the countryside did not happen. However, by 1912 the Government of India was facing growing political turbulence. The last decades of the nineteenth century and the first decades of the twentieth saw the telegraph world caught in a spiralling information panic that reached a climax with the First World War. This book suggests that this information panic was as much a symptom of technological change as it was the means for it. The inability of the telegraph to meet the financial, social and ideological challenges posed by newer technologies, and its resistance to change both financially and in terms of propaganda was reflected in the information panic, which signalled a world in transition. Ironically, the information panic illustrated both the ideal of perfectly stable information systems and the reality of their processes of change. From this point of view the panics were not irrational and unpredictable epiphenomena that suddenly swamped society but were inbuilt in ideological and temporal systems of information. The conflict between claims made by the telegraph in its late nineteenth century liberal and industrial ethos and the reality of its situation challenges simplistic notions of increased communication that leads 'naturally' to progress, free information, free markets, and peace.

In political terms, this representation of metropolitan control was wrong. Contrary to popular belief, the telegraph did not necessarily lead to greater control by Britain over the colonies. Rather, in some instances, it made Britain more a victim of events elsewhere and vulnerable to the decisions of the 'man on the spot'. Contemporary commentators noted this fact, as well as the increasing attempts by British parliamentarians to use telegraphically transmitted information to intervene directly in colonial issues. The *Times* complained of '... the disadvantage to India of the frequent Parliamentary interference which has been made easy by Sir John Pender's cables ... It serves to bring Indian policy more under the control of those who are ever on the watch for an opportunity of shaping it to their own ends.'[125] In contrast, the Viceroy Lord Minto noted in 1907, 'The more opportunities we throw open to the educated Indians [read nationalists], the more we shall have to increase the executive authority of the Viceroy and the Government of India.'[126]

Viscount Morley, the Secretary of State, claimed that a 'Secretary of State is a personage not to be trusted, that he is a puppet of Members of Parliament who, in their turn are the puppets of native wire-pullers.'[127] This was not a situation of control but a situation of loss of control where information often neither flowed into an imagined centre nor did commands necessarily flow out from it. The tensions between the India Office, the Government of India, the Provincial Governments and the man-on-the-spot continued.

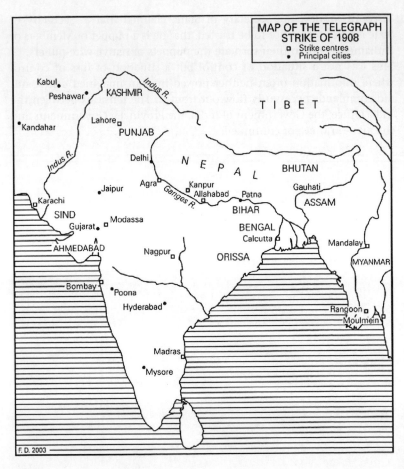

Map 7.1 A map of the telegraph strike

7
The Telegraph General Strike of 1908

Introduction

The period 1905–10 has been so far conventionally accepted as the Swadeshi and Boycott period and the first phase of revolutionary extremism, while it sometimes figures as a period constituting a prehistory of labour movements, popular politics, and communalism. This chapter investigates the relatively less studied telegraph strike of 1908. This strike occurred simultaneously in Rangoon, Moulmein, Calcutta, Allahabad, Agra, Bombay and Karachi. Both telegraph signallers and subordinate staff went on strike in these places. The strike was underway when the first revolutionary extremists were arrested in Calcutta. The entire focus of the government and the India Office in London was on this anarchist threat. Through a study of the information panic that ensued, this chapter examines these two events to argue for particular, though not necessarily co-ordinated, investments by different sections of the state and media in the construction of a narrative of events.

Previous imperialist and nationalist historiography has subsumed this telegraph strike within the master narrative of political unrest under Swadeshi and Boycott. By investigating the strike in detail, this chapter questions this master narrative. In particular, it focuses on the romanticism inherent in the exaggerated importance accorded to the political general strike, such as the Tilak strike of 1908 in Bombay. Returning to the work of G. D. H. Cole, a contemporary of the period, this chapter argues for the importance of a transregional and transhierarchical industrial general strike in the history of labour movements. By concentrating on the relationship between technological change and workers it goes beyond questions of continuity and representation, and demonstrates how workers across this part of the Empire, and globally, were capable of charting general agendas in the first decades of the twentieth century, using technology to combine and combat technological rationalisation.

The telegraph strike of 1908, which occurred at many nodes of the telegraph system in British India and Burma, was an important event at the time. However, it has not been dealt with to any extent by the histories of the period. The strike occurred among different sections of workers in Rangoon, Moulmein, Calcutta, Agra, Allahabad, Bombay, Madras, Lahore and Karachi. The reappraisal of the strike and the form it took against the political contour of the time allows us to recover social and political tendencies that have been pushed to the margins in the historiography. The strike was a protest action by the first virtual community, which was formed by the workers within the communications industry. The limits and potential of cross-class, transnational and cross-racial co-operation, and the politics of identity and representation, are the central foci of our study of protest by this virtual community. The chapter reassesses recent writing on strikes to argue for the need for a typology of strikes along the lines suggested by G. D. H. Cole's work in the first half of the twentieth century. It proposes the category of the trans-regional economic strike as an important tool in analysing strikes in the communications sector.

Previous historiography has subsumed the telegraph strike within the master narrative of political unrest under the Swadeshi and Boycott movement and historians have largely ignored strikes in the communications sector by various classes of workers.[1] These studies conventionally begin from the 1880s and characterise the period 1880–1919 as the 'prehistory' of labour mobilisation in India[2] and as the period of the emergence of 'community consciousness' within Indian labour.[3] This chapter, by studying protest in one of the core sectors of industrial technology, questions this chronology and indicates certain unquestioned continuities that have continued in the writing of history from the period itself. The continuing study of relatively underdeveloped agro-economic sectors such as tea, cotton textile and jute has led to the argument of the peculiarity and backwardness of Indian labour: labour in these sectors reflects nothing but the backward forms of industry they are situated in. Class solidarities in this case were formed directly through the telegraph system, in contrast with the more inchoate, mediated and nationalised identities formulated in the work of Marx and Benedict Anderson. There were two broad categories of workers that went on strike in 1908: the subordinate sections of clerks and peons, usually Indians, and the signallers, of which 75 per cent were European and Eurasian.[4] This work suggests that community and race identities were not inherent but hardened through the experience of working class action and the different ways in which the colonial state responded to it.

Reforms in the Indian Telegraph Department

To reconstruct the immediate and the general context of the telegraph strike of 1908, the conditions specific to the Indian Telegraph Department and the changing structure of technology and politics within which it was situated need elaboration. A brief note was circulated between the highest levels of the Telegraph Department and the Government of India in 1904. It was a prelude to the more public Telegraph Committee of 1906 and envisioned a drastic reduction in the Telegraph Department to improve efficiency, to cut down on subordinate establishment costs, and to allow for increased automation. The report reflected the need of the Telegraph Department to compete with the cable companies and maintain parity in rates. The size and the cost of the establishment were not proving viable. This combined with the need to keep increasing the number of users by cutting rates.

The report proposed that an additional 20 per cent be added to the strength of the General Service of the Signalling branch through selection from within the service and that they would receive a higher rate of pay. The conditions of General Service being: (i) must be European and Eurasian, (ii) qualified in Code and Figure tests, (iii) liable for service in any circle, and (iv) reversal to Local Service was possible and permissible at any time.[5] They designed an 'overall reduction in establishment through new standards of work.' However, the authors of the scheme warned the government that it would take a 'considerable number of years before the disappearance of the existing signallers is affected.'[6] As an immediate recruitment and retention measure they proposed that in addition to pay, '... all European and Eurasian Signallers, who compose seventy five percent of the total establishment [the total being 2279 in British India and Burma in 1904], be granted free quarters or allowance in lieu.'[7] The 'decentralisation plan' would then lead to savings in the clerical establishment that would reduce the budget cost to Rs. 3249, a saving of Rs. 1412. Over time more than a few livelihoods, especially in the subordinate sections, were lost.[8]

As an immediate consequence, signallers and clerks, fearing for their future, began to form a union. The Telegraph Association was formed and the movement for a subordinate relief fund was started. Within a few months, a further aggravating factor arose. Alfred Newlands, the Traffic Manager brought from Britain to reform the Indian Telegraph

Department, submitted his proposal for a series of reforms that would fundamentally change the Department. There was already resentment in the Department over the appointment of an outsider to reform the Indian Telegraph Department. Along with technical changes and increased efficiency tests, he also proposed ten-hour to 12-hour shifts. The workers demanded five and a half hour shifts or a six-hour shift, or eight-hours with holidays. It should be remembered that they worked nights, weekends and got leave infrequently. Sporadic strikes began to break out in the public sector and in the mills.

The Telegraph Committee's report, which was submitted in 1907, had some further disturbing proposals. During its deliberations it received as many as 87 joint petitions from the lower classes of the establishment. To summarise the issues, clerks wished to be accorded the same status as signallers with openings for induction into the signalling establishment. A provident fund was applied for along with a proposal for the formation of an association for welfare and recreation. Better medical facilities or an allowance was sought. The prevailing system of financial penalties for mistakes was protested against, as were very long distance and frequent transfers. Finally, revision in the scales of pay, promotion, and the policy on temporary employment, were issues of importance. Apparently the department had been using casual labour and a clerk or peon could be classed as temporary for as long as 15 or 20 years with no prospect of getting a pension or benefits.[9] Nearly all the demands were refused. Attempts were made to amalgamate the post and telegraph departments, and substantial reductions in the clerical and delivery establishment were proposed.[10]

The Telegraph Committee proposed to freeze recruitment and induct women[11] and military signallers, Eurasian and European youths from orphanages and mission schools in India. They also recorded their frank reservations against 'smart' men of Calcutta, Bombay and Madras. These standard pools of successful examination recruits were to be replaced by women and young recruits from the orphanages. The argument against the Bengali, the Bombayite and the Madrasi was simply that they might not be physically in shape for the task of touring and inspecting offices. Arbitrary fines, penal transfers and temporary, unpensioned and insecure employment were genuine grievances and had been acknowledged as such by the Telegraph Committee which wrote, 'Not only is the organisation of the signalling establishment defective but the existing rates of pay are ... inadequate.'[12] The Committee recorded that between May and October of 1906 alone, as much as 18.5 per cent of the signalling establishment was transferred, often for very long distances. It stated that this percentage of transfer, '... practically amounts to the transfer of the whole staff of every office in three years.'[13]

The conditions for the employment of women were: (i) that they should be between 18 and 30 years of age; (ii) that they file a declaration signed by their parents or guardians or some 'responsible authority' that they would live with their parents, guardians, or at 'some recognised institution.' While candidates for training were found in Madras, in Calcutta, Bombay and Allahabad, this proved more difficult. The Secretaries of Young Women's Christian Associations and Girls' Friendly Societies, advised the government to lower the recruitment age to 16, since most girls left school at that age and quickly found employment. They also advised higher wages, a shorter probation period and less than seven-hour shifts. The government adopted these recommendations.[14] However, the immediate problem with the signallers remained unresolved.

Information anxieties of the Government of India

The government had two important concerns regarding information: there was a demand for information as well as a need to contain information. At both levels of information flow, outward from the government and inward to the government, the state was exhibiting a need to process and contain the flow of information. Situated within a massive economy of information, one of the main concerns of the government was to contain information and prevent media access to sensitive documents. The government threatened officials with penal proceedings for passing on secret information. The government also needed to check combinations among its employees. The Official Secrets Act of March 1904 tried to set up these barriers. Collective bargaining by its employees was prohibited. The international climate of workers' movements, unionisation, and association among various classes of workers contributed to the concerns of the government.

The Government of India, shortly after the passing of the Official Secrets Act, cut its rates on international and national telegrams and the rates charged on international press telegrams between 1904 and 1905. First, it needed to increase information flows to maintain its competitiveness in the international telegraph network dominated by the cable companies. They wanted to increase revenue by cutting rates to increase the volume of business and square their rates with the falling cost of the individual telegram because of growing automation within the telegraph industry. Second, it was an attempt by the government to know the currents of unrest that were suspected to flow beneath the surface of indigenous society. Printing was encouraged and the cost of registration of newspapers was lowered. A number of politically radical newspapers began to be published in different parts of Bengal. For example, the daily *Bandemataram* was taken over by Bandemataram Company of limited liability, which had on its list of promoters, socialist radicals, Congress extremists, and revolutionary or populist ideologues. The list of promoters included, Subodh Chandra Mallik, Aurobindo Ghose, Chittaranjan Das and Bipin Chandra Pal.[15] *Charumihir*, published in what was now after partition called Eastern Bengal and Assam, was a typical paper that condensed Calcutta happenings, had a group of correspondents and a rapidly growing readership in the Province. There was a rapid increase in demand, circulation and the speed at which information circulated. It was a very volatile situation.

The partition of Bengal and the fear of a second mutiny

The partition of Bengal in 1905 helped populist propaganda to focus on the government as a target for different protest actions. In the Punjab, the Colonies Settlement Act further aggravated the problem. Urban and workers unrest now spilled into the countryside where the literate middle landholder and service gentry combined with the rural elite and, in the first phase of the anti-Partition movement, Zamindars in east Bengal. A mix of government policy, itinerant preachers, revolutionary and extremist propaganda, situations of famine, disease and discontent were bringing different strands and networks of discontent together. These waves were occurring in a situation of famine in eastern India and the plague in western India, and a sharp rise in the price index and in the general cost of living.[16] The Durbar celebrations held by Lord Curzon commemorated, as a side event, the fiftieth anniversary of the victory of the British in the mutiny and uprisings of 1857. This propaganda was mirrored beyond the control of the government in the countryside where similarly celebratory rhetoric resurrected the spectre of 1857. The government fell victim to these whisperings as much as it had helped generate them by its commemorations. Astrologers, revolutionaries and government officials were now working on a common schedule in anticipation of an uprising around 1907-8.

Preparations for the telegraph strike

Henry Barton took the initiative for telegraph unionisation in the autumn of 1907. The Director General of the Telegraphs later explained

Barton's motives away as pique at a missed promotion in the report to the Viceroy on Barton.[17] He was a fairly senior man in the service having served 26 years, 11 months and 12 days before he resigned. It was noted that he 'did good work' in upper Burma which was officially recognised at the time.[18] However, this did not entitle him to any consideration in the eyes of the department. The Director General posted him from Rangoon, the capital of Burma, to take charge of the office in Berhampur. It was not a central location and Barton refused to go to Berhampur. Instead, he asked for a posting either in Rangoon or in Calcutta. When his request was turned down, Barton resigned in January 1908 and proceeded on a 'tour of the principal telegraph centres in India and delivered inflammatory speeches to the men. Owing to the fact that he possesses undoubted ability above the average of the ordinary signaller, and to the gift of delivering addresses, he has had great influence over almost the entire signalling staff.'[19] Henry Barton as Secretary to the Telegraph Association began to publish the *Telegraph Recorder* from Rangoon in January 1908.[20] It is perhaps this early revolt against arbitrary transfers that allowed the movement to gain in coherence and organisation.

From the start Barton was a problem for the government. He had none of the discretion essential in government service. In December 1907 or January 1908, Newlands met with Barton in Burma while Newlands was on tour as the Traffic Manager. Barton brought along the local representative of the Telegraph Association to this meeting who was on the staff of the *Rangoon Times*. The details of the interview were published in both the *Telegraph Recorder* and the *Rangoon Times*. A lecture delivered before the staff by Newlands was similarly cited in detail in the same week in these publications. The Director General, in his note to the government, saw this as a clear breach of official etiquette if not actually of the Official Secrets Act. He held that Barton 'was mainly responsible for the appearance of the communications.' Moreover Barton in his February 1908 address in Calcutta, 'in contravention of the special orders, … read out an official communication which I had issued to the staff, and which was in consequence published next morning in the local papers.'[21]

The telegraph workers movement started with fresh impetus in November 1907 as rumours about the impending submission of the Committee's report began to circulate. Most of the establishment suspected the report to be unfavourable. It was reported in the press that 'messages were exchanged between the Signallers in all the main offices in India and a general assent was obtained to concerted action … great excitement at Bombay, Calcutta, Karachi, Madras, Rangoon, Gauhati, Allahabad and other offices.'[22] This movement was a process by which anonymous undistinguished signallers tapping telegraph keys became individual identities in order to coordinate and communicate between each other. The telegraph workers had to engineer a crisis of sufficient dimension to escape victimisation and to underline their irreplaceability to the state and business. The volume of messages between telegraphers was growing and a different network and virtual community with its own politics had begun to function within the telegraph network. Rangoon was one of the main coordination centres for the movement. Other important members of the workers movement were D'Rozario, Lafound, Isaac and Moses. In December 1907, the Director of Criminal Intelligence reported on the agitation among the subordinate staff and the activities of the representatives of the Telegraph Memorial Committee in Rangoon.[23]

By December 1907 Henry Barton was addressing telegrams on behalf of the Telegraph Association.[24] These telegrams requested a reply to the general memorial and listed their grievances, which included a protest against overtime. Above all they demanded the early publication of the Telegraph Committee's Report submitted in December 1907. The flows and eddies in information supplies crucially determined the chronology of events. The government declined to publish the report. The *Times of India* complained that 'although the Committee's report was delivered at Simla a year ago, the public know nothing of it.'[25]

In February 1908, the entire staff sent identical memorials to the Viceroy. The government in reply stated that it adhered to its decision 'not to publish the report of the Telegraph Committee until they have submitted their recommendations to the Secretary of State and his orders obtained there on ... Regarding the questions raised (by the petitioning workers) ... the Government are at present unable to hold out any hope.'[26] The Government of India was extremely sensitive to the press and had to be very aware of the language of their official publications, especially in the context of the growing heat generated in the British parliament on Indian issues and with the subsequent election of a Liberal government in Britain. The Telegraph Committee's report could not be published in the form in which it was submitted. The telegraph employees of Allahabad met on the 19 January 1908 to agree to join the Subordinate Relief Fund, and significantly, 'promising to enlist all absentees in the same cause'.[27]

The flooding of the government with petitions

The Government of India was surprised by the sheer volume of messages that it was inundated with: 116 identical memorials were sent to the Viceroy on one day from the signal room clerks in the Bombay division alone.[28] Many sections among the clerks, signallers, and peons coordinated to achieve this effect. Waves of petition followed with growing concentrations from December 1906 up to February 1907.[29] In January the entire signalling staff of India and Burma sent in almost identical petitions to the Viceroy.[30] The government was also hit by the fact that the petitions poured in from different parts of the system. Nagpur, Bombay and Karachi clerks experimentally coordinated to be received on the same day. Nagpur sent its petitions on the 8 February[31] as did Karachi[32] while Bombay had sent theirs two days earlier on the 6 February.[33] An infuriated Director General lashed out in his report to the government about the problem of 'surplusage of temporary clerks' and the identical nature of the mass of submissions, 'The generality of the prayers made are such as to court refusal.'[34] The Director General had in no uncertain terms pointed out, 'as a rule they [the clerks] are not up to the required standard of education, and it is most undesirable that a clerical post should ever get to be looked upon as a stepping stone to the Signalling Establishment.'[35]

The strike of the peons

On Thursday night, 27 February 1908, the delivery peons of the telegraph department went on strike. There had being growing discontent as new methods of delivery and attendance were introduced over December and January. Barton, General Secretary of the Indian Telegraph Association, asked for public sympathy and support from the press.[36] He urged the workers to adopt constitutional means of action.[37] By Saturday the entire system of delivery in Calcutta appeared to be under threat. Initially 173 permanent and 193 temporary men went on strike. Their numbers swelled and almost 400 men were involved. The striking workers met at the Calcutta maidan, beneath the Ochterlony monument and held a meeting adopting resolutions. This echoed the methods adopted by the political parties across the spectrum. It was as much a protest as it was a publicity spectacle; the striking workers chose one of the oldest and most familiar locations for political speeches and meetings. They demanded the same wages as the Bombay staff, better hours and conditions of work, winter clothing, *batta*, and promotion according to seniority and, most provocatively, the reinstatement of the two peons dismissed from service as the ringleaders of the 1907 strike in Bombay.[38]

That there were wider issues and feelings involved is shown by the fact that by Monday, 2 March, the boy peons at the Calcutta Central

Telegraph Office, numbering about a 100, joined the strike.[39] Madras telegraph peons, numbering around 60, went on strike on 4 March.[40] Telegraph delivery peons in Bombay followed suit on 29 March in spite of the concessions they already enjoyed because of the strike in 1907.[41] The Post Office clerks sent a petition threatening to join in the strike.[42] The postal workers in Modassa, Ahmedabad, joined in the movement.[43] The same day, the clerks at the Accountant General's Office threatened to strike.[44] Everywhere around them there were workers striking: in mills in Tuticorin, 2000 men in the marine dockyard in Khidirpur in Calcutta, in jute mills in Chandernagar, 6000 workers at the railway workshop at Parel in Bombay. The Railway Mail Service was dismantling the railway and mail schedules. Letters were written to the press against the actions of the Inspector General, Railway Mail Service.[45] The labourers employed to look after the overhead lines in the Calcutta Tramway Company went on strike and were dismissed overnight.[46] In short, in 1907–8 many of the branches of the administration seemed to be on the verge of open revolt. The context was a huge rise in prices because of the famine in Bengal, Bihar and Orissa in 1905–6[47] and the failure of monsoons in 1907.[48] The clerical strikes ended suddenly: the boy peons and the delivery peons were summarily sacked,[49] and the 10th Jat regiment was deployed for the delivery of messages till a new establishment was in operation.[50] Peons were brought in from Jullunder and Delhi.[51] The striking workers showed precocity in their demands, organisation and solidarity in different parts of the country. Their method and strategies were mature. For example, they sent a petition to the Commissioner of Police requesting him to intercede with the Director General and the Superintendent of the Telegraph Department on their behalf or accept their resignations. Here they stressed the loyal and peaceful nature of their rally. They 'wished to hand over their badges and uniforms to the Commissioner of Police and requested him to get their dues from the Department.'[52] Their summary dismissal took the wind out of their sails, and they turned to the nationalist leaders in the legal profession to plead their cause.

Intervention by the middle class political leadership

A. C. Bannerji, leading member of the Bar and a prominent swadeshi leader, offered to help their cause. Bannerji wrote to the Deputy Director of the Telegraphs requesting an interview for himself and a couple of his friends to 'help you to bring about a compromise'. This was because 'they (the workers) cannot by reason of their ignorance represent their grievances clearly'. He went on to write that it was a 'great pity that the claims of these poor people, the indigent children of the soil, should have been so despised ... I have no doubt that there are kind hearted gentlemen in every community who will take up the cause of the poor peons and try to set things right for them.'[53] Thus, they were now represented as voiceless 'poor wretches'. The *Marattha* and Gokhale spoke out on their behalf.[54] The *Gujarati* wrote:

> [P]oor Indians have become experts in the art of handling this weapon [strikes] imported from the west ... a good omen for the political future of India. The authorities and the Anglo-Indians seem to believe that it is the educated Indians that incite the labouring classes ... it seems to be their opinion that a subject nation is not entitled to have recourse to strikes, which are regarded as the exclusive property of European workmen. ... The resolution of the Telegraph peons to turn to other means of livelihood confirms our belief that their cause was not a bad one, and that their successors will reap the benefits of their agitation.[55]

The peons showed both organisation and courage. The workers combined in Calcutta, Bombay, Karachi and Madras to go on strike on both general and specific demands. These were primarily Indian workers. Their willingness to sacrifice their jobs showed increasing politicisation in their ranks. By the time of the Tilak General Strike their political awareness was considerably higher.

The strike of the signallers: 'Passive resistance'

Alfred Newland, the Traffic Manager imported from Britain launched a scheme of eight-hour work shifts, implemented in all offices on 3 April 1908.[56] The *Statesman* reported a heavy accumulation of messages in Burma while the *Bandemataram* discussed the warning issued by Director General Berrington which pointed out that the rate of sending is being purposely and wilfully slowed down and that '... faults on wires and apparatus are also abnormally high.' It was reported that 'large numbers of men in Rangoon and Mandalay had reported sick'.[57] Throughout March the government had been occupied with the dismissal and re-employment of the delivery establishment at some of the vital centres of commerce and communication. The last batch of dismissed Bombay peons handed in their uniforms and collected their dues on 6 April.[58]

By 8 April 1908 the system was in the throes of the second crisis. The main wires were fused and rendered dysfunctional. Engineering electrical faults or literally fusing the wires did this. On 6 April, 15 main telegraph wires went out of order in Calcutta. Bombay had huge accumulations of messages and subsequent volume of messages over the two telegraph lines handed over to operators of the Eastern Cable Company demonstrated that the signallers were deliberately slowing their output. The Calcutta Office staff insisted that it was Rangoon and Bombay that were holding them up. Rangoon was commonly believed to be the source of the trouble.[59] It was reported that Madras was coping but Bombay, Calcutta, Rangoon, Agra and Karachi were 'affected'. Lahore soon joined their ranks.[60] *Empire* thundered, 'the operators are fooling ... They waste their time in keeping the offices informed of the accumulation of messages, discussing the situation and sending wires at different centres giving accounts of the Press attitude.'[61] The *Rast Goftar* joined in the criticism.[62] Tilak's *Kesari* wrote that while the 'signallers who had gone on strike are mostly Eurasian and European ... they have our sympathy. It is a little curious, however, that strikes undertaken by white employees are always successful, while those organised by native subordinates fall through.'[63] Another main wire was found fused in the Calcutta Office and by 10 April, 12 noon, around 8000 messages were in delay and the figure was growing.

The Director General issued a fresh circular reminding the staff of the circular of 22 February and demanded 'loyalty and good sense' from them. Superintendents of Rangoon, Calcutta, Bombay, Agra, Lahore, Karachi and other offices facing similar problems were given the power to dismiss arbitrarily up to ten per cent of the signalling staff. Postal signallers would be supplied as replacements.[64] The Bengal Chamber of Commerce appealed to the Viceroy on 7 April to appoint a Conciliation Board as had been done in the case of the railways. The Viceroy's personal secretary telegraphed a reply that refused Viceregal intervention and was virtually a tirade against the subordinate staff that had 'chosen to deliberately block the introduction of the new hours of duty by delaying messages and absenting themselves from duty.'[65]

The workers used what they and the government called the strategy of 'passive resistance'. Slowing down of work speeds, engineering faults and collective absence at work through medical certificates and other forms of legal leave led to pile-ups on most of the major lines. Pile-ups meant delays not immediate disruption. Delay meant that the unified world time and the emerging global market would be in jeopardy. One of the primary difficulties faced by the administrators was the fact that this was a system of electronic communication. It was more concerned with flows and motion than with posts and stations. The workers could delay the system considerably yet blame the next station and it was difficult in an emergency to pinpoint the exact source of the accumulating information snowball: delay, in this case, was cumulative. This element of surprise would be removed when the workers went public with their demands. The realisation of this second strategy dawned on the government slowly. It became aware that there were too many accidental breakdowns, absences and pile-ups to be a coincidence.

The course of the signallers' strike

Henry Barton, secretary to the as yet unrecognised Telegraph Association, addressed a large meeting in Rangoon. The Director General of the Indian Telegraphs' threatening circular in response was deemed an 'egregious blunder' by the *Rangoon Times*. The Director General's circular prohibited any further use of the workers' club premises for meetings and threatened dire punishment for attendance 'where such language was used' and warned the signalling staff that 'unauthorised publication of information obtained officially' would render them liable to prosecution.[66] This was reported by the local press along with the government's refusal to countenance a provident fund for the subordinates, arguing that it was too close a copy of the privileges of the superior establishment.[67] A number of letters began to appear in newspapers complaining about the plight of the clerks, signallers and delivery service, citing unhappy working conditions as the reason for the prevailing mismanagement.[68] The Bengal Chamber of Commerce and the Calcutta Trades Association voiced criticism and concern.[69] The Director General of the Telegraph, T. D. Berrington, agreed to meet the representatives of the Chamber of Commerce in secret: '[T]he Director General stipulates that no reporter are to be present and that no report of the meeting will be published in the newspapers.' The negotiations were conducted over telephone![70]

As soon as the government's reply reached the staff through the Chamber of Commerce, Rangoon declared themselves on strike. Discontent was reported among the signallers in Chittagong.[71] The *Statesman* reprinted a telegram from Barton announcing the strike in protest against the summary dismissals from the staff at Rangoon.[72] The Director General summoned Barton from Rangoon to Calcutta. Upon his landing, Barton addressed a large gathering in the town hall. Again this was a political choice, aiming to maximise visibility and media coverage. The signallers went on strike in shifts so as not to lose out on their day's pay. The signal 'Diabolic 15' was flashed to all the offices. It meant 'general strike at three p.m.' There is a detailed description of the strike in Calcutta in the *Bandemataram*. The first batch went on strike in Calcutta at three in the afternoon. '... The new watch and the old watch gathered on the steps and beckoned those inside to come out by waving sticks, hats and handkerchiefs. Two Eurasian youths, no doubt in their excitement, yelled the fatal word "strike".'[73] In Bombay, the strike began at two p.m. Agra joined at four. Kanpur and Allahabad were rumoured to be ready to strike at five in the evening. In fact, Allahabad joined the following day. In Bombay, 25 signallers were served with notices of summary dismissal. The notices were dated 8 April 1908. In short, there seemed a genuine possibility of a 'general strike by the signallers throughout the country.'[74] Asked by the *Bandemataram* reporter whether the strike would be 'universal', a spokesman for the strikers said that some of the senior men, while in complete sympathy, would not join the strike and that these men would be of use in keeping open communication between the different centres. He added that the 60 Bengali signallers, out of a total of 240 in Calcutta, would not 'by any chance go on strike.'[75] By December 1907 reports had circulated in the press regarding discontent among Indian signallers who were after 1904 no longer encouraged to join the 'General Service' with better pay and prospects. It was alleged that the 'Department was closing its doors to Indians.'[76] Possibly, they could not risk any overt involvement, and in any case telegraph offices had to be kept open to co-ordinate this all-India strike. This lack of overt involvement within a section of signallers was hardly a threat to the success of the signallers' strike.

Emergency measures and panic: The omission of date and time on telegrams

As early as February, the Bengal Chamber of Commerce had warned the government, referring to the discontent in the subordinate ranks and had suggested 'an enquiry into their [workers] alleged grievances with a view to avert a strike and the consequent disorganisation of public business.'[77] The officials thought this blatant impertinence on the part of the Chamber. In his note on the letter, G. Rainy wrote that the Chamber's suggestion was 'quite unreasonable and unintelligent ... They are not in a position to advise Government. Their action can only be described as most unfortunate and tending directly to the encouragement of insubordination amongst the men.'[78] The Viceroy,

Lord Minto commented that '[T]he Chamber of Commerce letter was most ill-judged and unjustifiable.'[79] In their reply to the Chamber's letter, the government stated that the difficulties were being removed firstly by introducing reforms of a radical nature into the workings of the department, and secondly by improving the conditions of service of the subordinate staff. The government also categorically stated that they did not foresee a general strike of the signallers.[80] The Marwari Chamber of Commerce received a much more terse reply to their letter of 10 March.[81] They had complained of the 'Indian merchants and traders who have suffered and are still suffering considerable loss and inconvenience through the phenomenal delay because of the strike of the delivery peons.'[82] Even the Anglo-Indian papers were quick to point out that 'Trade and commerce everywhere throughout India is in a state of paralysis, and our happy-go-lucky Viceroy is away enjoying himself shooting tigers, apparently not caring a tupenny damn whether the commerce of India goes to the devil or not.'[83]

Large batches of postal signallers and military telegraphers were deployed to replace the striking workers. Burma was particularly hard hit because it had very few postal or military reserves. The main line often could not be operated. Another crucial problem involved the Baudot signalling instruments recently introduced on the main routes and in the main centres. The substitute signallers, so technologically out of touch that some of them had not tapped a key in the past three years, were rarely found to possess any working knowledge of the sophisticated high speed Baudot.[84] As a result, replacement of the strikers by sufficient numbers of efficient workers proved impossible. The Superintendent of the Agra office was removed. The reason being that he had panicked and contacted the Deputy Superintendent of Police who reported that

> the unrest among the telegraph clerks has extended to Agra, which is the biggest telegraph office after the three great centres of Calcutta, Madras and Bombay ... However, Mr Morgan [Superintendent, Telegraphs] was unnecessarily alarmed. ... *I send this by letter, and have not wired, as it is not desirable to attract more attention to the matter than is absolutely necessary and which would be the case if I sent long cypher wires under the present circumstances through the local telegraph offices* [emphasis mine]. ... The attitude taken up is obviously one of passive resistance and has taken the form of getting 'sick' ... at present the strike is not apparent to anybody unconnected with the with the Department.

He went on to add that Morgan had 'panicked.'[85] The quandary of the government having to avoid its main means of communication multiplied its feeling of panic and vulnerability.

The Telegraph Department resorted to the omission of the date of despatch from the telegrams so now receivers could not be sure of the time and date of despatch. This compounded the confusion.[86] The *Indu Prakash* wrote in its columns, 'We have before us a telegraphic press message, which bears neither the date of despatch nor the timing. While we can suppose that hours and minutes have been ignored in the hurry and confusion obtaining at present, it is difficult to ascertain why there is no room for the *date* of the message ... And in all business matters a good deal depends on knowing the date.'[87] 'To all appearances there is a complete breakdown in the telegraph service between Calcutta and Karachi and a partial breakdown between Bombay and Karachi. Telegrams are filtering through slowly and, in the absence of any date of despatch, it is impossible to say whether the messages were despatched on the day of receipt or on the previous day or a week before that.'[88] Similarly, 'at all the main centres efforts are being made to show a clean slate ... sub-offices are being shut down and men are being drafted in post-haste from the *mofussil*.'[89] The disappearance of date and time of despatch from telegrams seemed to threaten the very basis of the need for rapid communication. Marwari merchants were complaining bitterly at the delay in the opium despatches. None of the ordinary classes of telegrams were being accepted and the telegraph offices were accepting only the very urgent ones. The Government of India was forced to issue a notice in the government *Telegraph Gazette*, refusing to accept ordinary and deferred telegrams at their telegraph offices until further notice.

Denouement: Official recognition of the Telegraph Association

Henry Barton was again summoned to Calcutta from Rangoon. W. L. Harvey, Secretary in the Department of Commerce and Industry, wanted to officially meet with Barton, Secretary to the Indian Telegraph Association. This was the political breakthrough that the strikes had aimed for. Barton met with Harvey as the representative of an officially recognised organisation and 'an amicable resolution was expected.'[90]

The strike ended 12 days after it had started. Speaking to the Rangoon Telegraph Association in 1909, Barton described the 'history and spread of the Association during the past eighteen months' and advocated as their motto 'Definite Forwardness'.[91] On 22 April 1908 wires were sent to all centres in Burma and India announcing the return to work. Barton expressed his gratitude for the support of the press and the Chambers of Commerce. It was noted in the press, perhaps with pride, that this was remarkable, especially for Bombay which had been on strike for a record 12 days. The previous record had been eight days in the American signalling strike of 1907.[92]

The experience of 1908 and after showed how a very high degree of organisation and mutual sympathy could crumble without a platform. The Telegraph Association was both a symbol and a platform for unity for the signallers, and in this, the Chambers of Commerce helped the signallers. The peons and clerks had no similar organisation that could give coherence to a sustained campaign. However, this victory of the signallers had, as its underbelly, the sharp polarisation between the Indian and the Eurasian and European workers. Indian workers were increasingly becoming politically conscious and expressed themselves through mass strikes and resignations. The handing over of their uniforms by the peons revealed a particular bitterness as these uniforms were their badges of respect and the mark of their distinction in the community because of government employment: it was an extremely political act. Replacing them did nothing but swell the ranks of the growing 'nondescript crowd' watching events. The dismissed peons returning to their villages in the countryside and taking up other employment meant the spread of a greater commitment to political change and action. These experiences allowed the political process and the system of government to be understood. In contrast, the European and Eurasian workers had to be extremely unequivocal about their patriotism and loyalty to the Empire, and had to promise not to strike.

The telegraph strike over the summer of 1908 reveals that there was more than one strike in terms of different groups of workers and the chronology of their strikes, for example, the delivery peons, followed by clerks, and then signallers. Yet there was a perception of unity as they went on strike at around the same time in different centres. There was a developed awareness of the importance of propaganda and publicity for both classes, and, especially in the case of the signallers, there was a high level of co-ordinated action. The strike was not a millenarian or communitarian uprising but one that was integrally linked to the world economy. It was the direct experience of state repression and representational politics of the time that led to the subsequent polarisation and hardening of community identities among the workers. The Eurasian and European signallers were re-employed while they agreed to try out the new working system and hours. Though many requests were made to re-employ the disbanded peons they were not take back.[93]

The workers struck work over the government's refusal to publish the Telegraph Report. A paranoid bureaucracy, obsessed with secrecy, sedition, surveillance and policing generated its own scares and panics. A strike at the centre of the communication system increased its sense of panic. It could not communicate while using the system that was in revolt. This was the second mutiny the government had been so anxious to uncover, yet seemed completely surprised and paralysed by it when it happened.

Should these strikes within the same sector be studied separately or can there be such a thing called the telegraph 'General' strike of 1908? E. J. Perry distinguishes between waves of strikes that 'develop around work-related grievances ... criticism aimed at factory management' and a general strike, which is characterised by 'broad political demands that stimulate extensive cross-class enthusiasm targeted directly at the state.'[94] She cites the example of Russia starting with the St Petersburg cotton spinners strikes of 1896–7, metal workers with socialist ideas in 1901 and the general strike of 1905.[95] For her then, the General Strike is a telos to which other strikes necessarily build up to. She emphasises the workers subordinating work-place grievances 'in favour of assuming a role as citizens ... It was only when students managed to reach out to allies among the working class, that their quest for citizenship ... took on exceptional political force.'[96] Perry is enthusiastic about the potential of a 'general social strike' and the progression of the worker from civil citizenship through political citizenship to social citizenship.[97] G. D. H. Cole distinguished between different types of 'general strikes'.[98] There was the 'Political General Strike', which aims at a specific reform that it calls on the legislature to pass into law.[99] There was the 'Economic General Strike', devoted to gaining some definite concession from the employer.[100] There was the 'Social General Strike', aimed at the complete overthrow of capitalist society and the substitution of a new order.[101] Finally, there was the 'Anti-militaristic or General Strike against war.'[102] Writing in 1913 he saw little future in Britain of a 'Social General strike'. 'The General Strike ... is anarchist in its origin, and has throughout the unpractical and Utopian character of Anarchist ideas in a very marked degree.' This was especially true of the 'anti-militaristic' and the 'Social' General Strike.[103] This classification serves as a useful point to enter into the question of the nature and character of the telegraph strike of 1908.

What is clear in the context of the telegraph strike of 1908 is that an industry-wide strike was effected involving vertical as well as lateral co-operation, that is peons and signallers had sympathies with each other and they were striking at the same time in different centres. Applying Perry's model to India would mean a progression from civil to national to a worker conscious of social security and welfare issues. This was something that contemporary nationalists used in a sense when they argued for *swaraj* above and before everything else. Nationalist historiography of this period has tended to follow a similar trajectory in tracing what might be called the 'evolution of notions of citizenship.' Nationalist mobilisation and student activists, it is argued, gave workers movements the necessary political aim and coherence; they helped create a politically mobilised working class. A typical example of such unionisation would be the following advertisement:

> The Railway Union. ... Each person on being a member signs a pledge, and takes a vow to further the cause of brotherhood faithfully and loyally and to submit to all rules and orders of the Union. Unions are being formed in the districts and local centres. An organisation is on foot in Calcutta to form a Union of all these unions. Secretaries of local or branch unions are requested to send without delay the names of their office bearers and committee and the total number of members.

The letters were to be sent to A. K. Ghose, Bar Library, Calcutta High Court or 39 Lower Circular Road, Calcutta.[104] The private address was of a wealthy lawyer and doctor enclave while the High Court address emphasised Ghose's middleclass respectability. This middleclass intervention in the Railway Union has been overextended to suggest that they were a key element in working class mobilisation in this period. The unionisation that occurred under the patronage of the intelligentsia was, as the quote above illustrates, often a very Masonic affair.

Aetiology of the strike

From the time of the strikes, reports had circulated explaining the cause of the strike. It was a conflict over representation, and the middle class media rallied together in their attempts to explain and co-opt working class protest. The Swadeshi movement, activities of early lawyer trade unionists, and international workers strikes were some of the explanations offered. The nationalist *Kal* wrote, 'A strike is an indication of public discontent. ... It is said that the cause for discontent is their meagre salary. But this was not so. The wave of patriotism flowing all over the country seems to have touched them [the workers] also. ... Their demand is for Swaraj.'[105] The *Bandemataram* reprinted an article from the *Induprakash* asking, 'Are they [strikes] not offshoots of the general movement of boycott initiated in Bengal? Similar strikes are taking place everywhere and in various departments. Recently there were strikes of ghariwallas in Ahmedabad and of the native staff on the East India. Just now there are strikes by scavengers and mill hands in Calcutta ... and strikes are threatened in the Police and telegraph Departments ... are they not a new awakening of national life?'[106] Others spoke of the recently imported fashion of strikes; something learnt from the western workmen and alien to India.[107] They were a spin-off of the 'mania for striking, which seems to have become epidemic in all the big towns of India.'[108]

Henry Barton, Secretary of the Telegraph Association, was forced to publicly protest that the 'movement was free from anything approaching insubordination or disloyalty'.[109] The agenda of the worker was in retrospect classified as 'political' unrest by collusion between the authorities, union leaders, and the middle class elite. The strike of 1908 became a victory for nationalist leaders capable of mobilising working class sympathies and a prologue to the Tilak political general strike, an event in the history of the Swadeshi and Boycott movement. The workers themselves, through the unionisation, realised their union while sacrificing any more general political and economic combinations and objectives. A fresh generation of activists was indeed involved with the labour movement between 1905 and 1912 but these activists were neither systematic nor particularly well organised. In the case of the telegraph strike their roles were more as spokespersons for the workers than as organisers or mobilisers.

The political general strike of 1908 in Bombay over Tilak's arrest and deportation has gripped the imagination of historians who see this period in terms of early efforts at mobilisation of labour by nationalists.[110] Many different strikes and trajectories have been fused together to lead to the Tilak General Strike of 1908.[111] The prominence given to the Tilak general Strike of 1908 as both a telos, which all other workers' unrest was leading to, and an event of significance in the process of nationalisation of workers politics, stems from collapsing the social and political general strike with the economic strike. There are two features of the social and political general strike that are noted. The first is their radical social and political content that allowed cross-class co-operation, and nationalist mobilisation. The second, perhaps less obviously, is their localisation in practice. The Tilak strike was general in content and mobilisation but was, by and large, geographically local: the Tilak strike was powerful only in Bombay. The study of general strikes in the social and political sense has also led to a concentration on the spaces and associations of mobilisation that included teahouses, messes, native place and caste associations and the *mohalla* or neighbourhood.[112] However, these networks were not primary or crucial in the case of the telegraph strike. Combining across regions and cities meant the neighbourhood had little significance. It is also important to realise that workers combined across differences of caste class, religion, community, and region in this sector of the industrial economy. For example, signaller M. E. D'Rozario, later one of the leaders of the strike movement, petitioned the government along with Dalloo, Sub-Inspector, Nagpur Division, and Linesmen Garubu, Construction Division, and KhodaDeen, Bengal Division, for extra pay while stationed at Aden.[113] This denies the contention that the Indian labourer was primarily a peasant and was cleft along communal and community lines.

Alongside the specific demands of the workers there were more universal and general demands. This echoed labour movements across the world. The universal standardisation of time and the increase in working hours reflected an international concern of labour. The first national strike or lockout in Britain had stretched for six months between July 1897 and January 1898 and a General Federation of Trade Unions was experimentally established in 1899. By 1908 a Labour Party was in existence.[114] In 1907 the telegraph workers struck work in the US. The Indian press reported in August 1907 that there was a 'great strike of telegraphists ... over 1600 operators in Chicago, sympathetic strikes in Denver, Colorado, and Salt Lake City ... The telegraphists strike has spread to 50 cities in the western and southern States.'[115] It was soon reported that Toronto and Montreal had joined in the strike and that 'communication has stopped throughout the USA except by telephone.'[116]

The press equally meticulously recorded allegedly similar kinds of labour and other unrest in Liverpool, Belfast, Egypt, France and Russia. This chapter suggests that there may have been a possible moment in the first decade of the twentieth century when workers might have combined across national boundaries. Historians have tended to concentrate on the political and social general strike within a city or industrial region but these apparently spontaneous political general strikes did little for workers in specific industries and were perhaps more important in raising the general political and social consciousness of the workers. The economic transregional strike, in contrast, was a general workers strike and successfully held the attention of business, state, and the media, simultaneously across several centres. It worked as a powerful tool to wrest specific gains for the workers rather than as a part of a political movement. In the end it was both a failure of representation and an inability to avoid unionisation that led to the abandonment of broader political ambitions and the cross-class co-operation of the strike. The emergence of community and racial identity in the process of working class unionisation reflected on a smaller scale the partial democracy, the form of political or rights representation, and the communal award system soon to be introduced at an all-India level.

Conclusion

The telegraph strike did not occur in an historical and geographical vacuum; it was neither isolated nor local. It was transnationally contiguous with other labour movements elsewhere in the world, particularly the US.[117] The strike forged links across caste, race, and, class boundaries to become a labour event, stretching across larger India, from Peshawar to Rangoon. This chapter discovers this all-India strike by a virtual community.

This chapter has questioned and revised several assumptions. First, that labour mobilisation and movements from the 1880s to 1919, and even later, constituted a 'pre-history' of labour mobilisation in India.[118] Second, that this labour in India was essentially made up of seasonal peasants and divided hierarchically along caste and community lines. Finally, these crucial mobilisations across class, caste, race, and regional boundaries in the telegraph strike were subsumed within a 'nationalist' and patriotic history. The end product of these diverse alignments across boundaries was until now believed to be leading to a climax: the very local and very nationalist Strike of Bal Gangadhar Tilak in Bombay.

8
Swadeshi and Information Panic: Functions and Malfunctions of the Information Order, c. 1900–12

Introduction

The *swadeshi* and boycott movement,[1] and the revolutionary movement were a part of the Information Empire in which the telegraph was an essential factor. The particular moment and nature of articulation of *swadeshi* was the product of rapid informational change. Telegraph rates and press rates were crucial to its rise. This chapter shows how imperialism created the technological environment which facilitated the growth of its opposite: nationalism. The telegraph strike was as much a political crisis for the government as was the *swadeshi* and subsequent revolutionary movements. Both were part of a global information order, though each claimed independence from each other. The international representation of Indian nationalism was followed by its articulation in India. The lowering of telegraph rates and the loosening of Reuters grip over Indian news facilitated this nationalism. For nearly a century British India was 'destined to play a central and highly profitable part in the Reuters empire within the British Empire'.[2] Reuters monopolised the supply of news to both English and vernacular papers[3] until further reductions in rates allowed more regional news reporting to emerge. Traditional historiography is discussed in this chapter to show how fresh insights can be gained by analysing communication systems and the information order.

The Government in India replaced the striking telegraph workers with operatives from the army. The 10 Jat regiment was brought in to resume telegraph traffic, but by the end of 1908 there was evidence that the seditionists had infiltrated the regiment. The regiment as a whole was not punished but merely sent back to its base in the northwest, and some of the men were discharged from service. These demobilised men returned home to their villages. The district authorities, Amraoti district, United Provinces, extended the Seditious Meetings Act, that had previously applied to Bengal, to this remote district in central India.[4] Soon after, the Government of India extended the provisions of the act throughout India. This particular trajectory of sedition that started with strikes in the government sector was not reflected in the reports sent back to parliament in London. It is a matter of significance that the government was faced with a revolutionary conspiracy almost at the same time.

In the early weeks of May 1908, the first revolutionary conspiracy was discovered. The telegraph strike had barely concluded and the workers were returning to work. A revolutionary attack went awry and brought back vivid memories of 1857; ordinary white women, children and men seemed threatened. Two young bombers sent to Muzaffarpur in Bihar from a district in Bengal, attempted to assassinate Justice Kingsford who had ordered the public whipping of schoolboys found to have been involved in *swadeshi* activities. They killed instead a mother and her daughter. Shortly after this, the Criminal Intelligence Department raided a garden house in a suburb of Calcutta. By the next week a conspiracy was unravelled starting in Muraripukur Garden, a suburb in Calcutta and stretching to Nasik, Pune, Paris, London, and Bombay. Networks of conspiracy were unearthed that led from Muraripukur to the India House founded by Shyamji Krishnavarma in London, and from there to Madame Bikaji Cama, a Parsi lady with anarchist links based in Paris. Paris was the centre for many of the malcontents in Europe and was reputed to be a training ground for revolutionaries. Bomb manuals were sent from Paris to Pondicherry, a French protectorate in India, and then to Calcutta and Pune. The gang in Alipur was also found to have established contacts with similar groups throughout the districts of Bengal and as far way as Nasik and Pune in western India. Reports of revolutionary organisations in America and Canada contributed to the picture of a global conspiracy. This chapter discusses the processes behind this particular construction of causality and event. It questions the previously unchallenged grand narrative of the Swadeshi and Boycott period and the first phase of revolutionary extremism between 1905 and 1912.

The information boom

As noticed earlier, the government cut the rates charged on internal and foreign telegrams in 1905 and the postal rates in 1904. In 1900–1 there were 32 million newspapers, in 1902–3 there were 32.5 million, in 1904–5 this number increased to 37 million and by 1905–6 there were 44 million newspapers in circulation in India. In 1900 there were 24,572 miles of rail carrying 43 million tonnes of goods, and 5.5 million inland telegraph messages. By 1910, there were 32,104 miles of railroad carrying 71 million tonnes of goods, and 11 million inland telegraphic messages.[5] It is against this backdrop of increasing velocity of circulation of information that this narrative unfolds. The government was trapped by the need to increase profits and square the rates with a decreasing unit cost of a telegram.[6] It was also caught in a frantic need to know the 'real opinion' of the country. The immediate reason for this need was that 1907–8 marked the 50th anniversary of the events of 1857–8. The press argued that the 'cheapest and vernacular newspapers' had the most to gain. The *Pioneer* noted the fact that 'the bulk of this literature is dangerous stuff … especially during the last two years … the newspapers most violent in their attacks upon Government have the largest circulation … there are serious drawbacks in the *pice* postage now in vogue for newspapers.'[7]

The historical context

Before entering into the details of the revolutionary movement it is important to reconstruct the context within which these events were constructed. The Curzon regime from 1899 to 1905 was remarkable for the many different areas of government it was trying to reform at the same time.[8] This attempted rationalisation of the administration sparked off reactions in very different areas and groups in India. This chapter discusses political events in North India from 1905 to 1912. The immediate context for the general unrest was the proposal to divide the administration of the large presidency of Bengal and create a new province of Eastern Bengal and Assam. The government also undertook a series of measures that affected land tenures in the Punjab. The partition of the Province of Bengal by Governor General Lord Curzon in 1905 sparked protests among many sections of the population. The movement found immediate echoes in popular and revolutionary militancy in Maharashtra and Bombay in western India. The repeal of the partition in 1912 has conventionally been taken as the end of this phase of politics. Historians have labelled the period 1905–12 as the *swadeshi* and Boycott period and of extremists within the Indian National Congress and the first phase of revolutionary extremism.[9]

Valentine Chirol, one of the first published authorities on the revolutionary movement in Bengal, began his narrative of the 'unrest' with the Chapekar brothers.[10] As a reaction to the plague measures in Bombay in 1897, Damodar and Balkrishna Chapekar built the Chapekar Association into a violent organisation. Damodar Chapekar assassinated Walter Charles Rand, the Plague Commissioner, on 22 June 1897. He was tried and convicted. But other members of the association continued the violence. Two unsuccessful attempts were made on the life of a police officer involved in the investigations and the two brothers who had turned informers were murdered. By 1908, the associations had become public in a wash of blood. A spin-off was the arrest and deportation of Balgangadhar Tilak for publishing seditious articles in the *Kesari*. This sparked a huge strike in Bombay City and in the mills. This was the anatomy of the subsequent Bengal violence.

According to Chirol's narrative, seditious articles in the press by prominent leaders inspired small and secret organisations based around physical training and gymnasia who mounted a militant defence of Hinduism. The initial violence was followed by acts of retribution often against a host of informers, approvers, witnesses, investigating officers and lawyers. The assassination of Commissioner Rand heralded the development of a revolutionary movement along religious fundamentalist lines, described as a Chitpavan Brahmin conspiracy. However, it was the partition of Bengal by Curzon in 1905 that turned revolutionary extremism along Narodnik and anarchist lines.[11] While the comparison and similarity between movements in Russia and India can be tenuous, it serves as an entry point into contemporary debates about society and development. The Masonic structure of these revolutionary societies was reflected in many parts of the world.

The *swadeshi* and Boycott movement in India and the revolutionary extremism that gathered momentum from the partition of Bengal by Lord Curzon in 1905 and the plague measures, particularly in Bombay, have been the main focus of the work done on this period. Early accounts looked to caste and religion as the key explanatory factors of the 'unrest'. Official narratives followed this Chirol narrative of caste, religion and revolutionary conspiracy.[12] The Rowlatt Committee Report spoke darkly of 'perversions' and secret rites of initiation. It spoke of religion turned upon itself and violence as a symptom of unhealthy minds: a 'single movement of perverted religion and equally perverted patriotism.'[13] Subsequent studies dealt exclusively with the policies and the administration of the state in this period without dealing with the ideologies and groups in opposition to it. Histories also dealt with the first phase of the growth of the Indian National Congress in the context of the struggle in local municipal bodies and the Legislative Council and the municipalities.[14] More recent histories use class as an analytical tool and concentrate on the revolutionaries.[15]

One of the underlying motifs in these writings is the failure of the movement both in terms of popular mobilisation and propaganda. By concentrating exclusively on particular groups and state policies, many of these histories tend to implicitly judge the processes in the making of the history of this period.[16] Historians have conventionally traced the actions of these political groups either in relation to state policies and policing or in the context of long term processes at work in Bengal and elsewhere. Histories have dealt with ideology while stressing long-term economic and social causes and concentrating on the role of the 'Moderate' and 'Extremist' factions of the Indian National Congress in this period, for example, in Bengal and Allahabad.[17]

Historians dealt with parts of the unrest in terms of the emergence of communalism[18] and working class politics.[19] Work on labour stressed the importance of industrial mill workers in the shaping of state policy.[20] The intelligence department of the state has been examined in the period 1904–24 to argue that the growth, budgets and numbers involved were not commensurate with the thesis of over-development of the colonial state in terms of intelligence and surveillance.[21] This is contrary to the argument put forward that the British state in India was tending towards a kind of a police state.[22] In contrast, the diplomatic fears and anxieties during the Great War were analysed to show how contemporary perceptions of both policing and the Indian revolutionary threat were greatly exaggerated, as compared with what can be observed in hindsight.[23] There was increasingly strident propaganda about the empire in this period as has been shown through the study of imperial propaganda.[24] The main argument of this chapter is that the Chirol narrative of 'unrest', which has been accepted by subsequent histories, obscures and subsumes a number of equally important events. This chapter suggests ways in which this period can be re-assessed and emphasises the need to examine how the narratives of *swadeshi*, Boycott and revolutionary extremism were constructed.

The sociology of the movement

The unit in Bengal was the *dal*; a small, close-knit group organised around the leader or *dada*. It was an almost religious bond, involving a very close relationship between the leader and the follower, and blind loyalty of the second towards the first.[25] In its aims and organisation it was necessarily local.[26] The context of the *dal* would be the *para* or neighbourhood. The mushrooming of the societies also indicated high levels of unemployment and under-employment of youth, as well as high concentrations of youths in urban hostels and messes. This meant several things. These people were potentially very mobile, returning to their homes and receiving guests from there. They were also very isolated in these urban centres, evolving high inter-personal loyalties as well as socially irresponsible behavioural patterns. They could be relatively easily recruited and made mobile. They were also relatively easy targets for police raids focused on the student.[27]

The relationship between the leader and his flock was pastoral. But this was possibly not a special feature of these urban and semi-urban societies in the early twentieth century. *Akharas* or physical culture societies centred on the *ustad* were a traditional aspect of Indian life. The *Guru-shishya* or the *Ustad-shagird* relationship was pastoral. The collective aspect of the societies can be over emphasised. At times the only common factor was the leader. Strangers, often with different levels of allegiance, different agendas and definite functions met on the eve of the 'action', and were often not known to each other before the violent incident.[28] The levels of anonymity would appear to have been very high by 1908, partly as a result of police crackdowns. Under these circumstances the secret societies had to evolve a community of rhetoric, fervour and cosmology. Personal and daily interaction was probably increasingly replaced by more long-term, long-distance and more veiled contact. One of the major spin-offs of the drive for secrecy was the evolution of a language of the initiated. A student, in urban spaces, could turn into a rabid fanatic. This split personality probably saw increasing self-expression through a secret self. Bankim Chandra Chattopadhyay's *Anandamath* had already laid down the lineaments of secret societies.[29] It could be argued that this 'othering' reflected the lack of any legitimate means available for self-expression, leading to the proliferation of secret selves.

Internationalism of nationalism

By September 1907, reports and predictions about revolutionary conspiracie s were being circulated in the press with regularity. Astrologer Babu Tarini Prasad of Bengal predicted in Benares in 1907 that the Pacific Ocean would be the theatre of a naval war. He also predicted war in Europe. Britain and Japan would enter into a contest over their naval strengths. The power of Japan was predicted to grow considerably, 'and Korea and China will be subservient to its wishes.'[30] As early as 1905,

Lala Lajpat Rai addressed a public meeting where he asked Indians to 'spread over France, Germany, Japan and Canada': 'you will find the world will quake before you.'[31] Headlines in New York flashed in 1906, 'England hunts Indian rebels who shipped 100,000 guns.'[32] By early 1908 the press was reporting that 'arms are being sent from America and secret societies of Indians are mushrooming there.' The editor went on to suggest that, 'If the rumours were believed by the American public it was simply because the wish was father to the thought.'[33] It was rumoured that 'firearms are being sent in sewing machines and bombs etc. in tins of condensed milk.'[34]

There were international movements and events that contributed to the situation of brewing unrest. The Japanese victory in the Russo-Japanese war in 1905, China and Ireland's boycott and nationalist movements, and reforms in Egypt combined to create the context of the *swadeshi* and boycott movement. Newspapers described in detail the methods used by Chinese and Irish nationalists and the strategies of boycott in use. The awareness of international dimensions and contemporary slogans provides the context for Aurobindo Ghose's exhortation on 7 June 1907. Aurobindo Ghose wrote, 'What India needs especially at this moment is the aggressive virtues, the spirit of soaring idealism, bold creation, fearless resistance, courageous attack; of the passive tamasic spirit of inertia we have already too much. We need to cultivate another training and temperament, another habit of mind. We would apply to the present situation the vigorous motto of Danton, that what we need, what we should learn above all things is to dare and again to dare and still to dare.'[35] It was perhaps not surprising that Barindra Kumar Ghose, Aurobindo's brother, was arrested as one of the main conspirators from Muraripukur. Aurobindo Ghose was also arrested.[36]

More immediately a series of international conferences illustrated the emergent global network. India was strategic to the empire and international commerce, and the Indian concern was to project their cause in international forums. Dadabhai Naoroji attended the International Socialist Conference at Amsterdam in 1905. He attended as the 'delegate of the nationalist party in India ... and thus India made her unexpected and startling entry on the political theatre of the world.'[37] Madame Bikaji Cama attended the International Socialist Conference held in Stuttgart, Germany, and delivered a speech on 22 August 1907:[38]

> I have come here to speak for the dumb millions of India who are going through terrible times under the English Capitalists and British Government ... You are discussing colonies all the time, but what

about Dependencies? ... Should there be such a word as Dependency in Socialism? Is this not the fight between class and class ... How can they [Indians] start socialism in that land where there is no liberty ... Do not misunderstand me, I have every sympathy with Russia and Poland ... but let me tell you that India's sufferings are greater ... you, Socialists, must say that you are against the bureaucrats and the English capitalists; say that you are for the suffering millions of Hindustan.[39]

Conservative newspapers cautioned, 'we do not question the sincerity of Mrs Cama, but we cannot help saying that she has mistaken the sphere of her activity. Let her mix among the people of Britain ... what can these Socialists of Europe do for us? Absolutely nothing. Their hands are full enough, and we are firmly convinced that neither their gospel nor their methods will ever do for India. Our salvation lies in our own hands and in the goodwill of England.'[40] The Congress of Anarchists was to be held in the same year in Amsterdam. Reports to be read at the conference were 'Anarchism and Trade Union', 'Organisation of Anarchists', 'Political General Strike' and 'Alcoholism and Anarchism'.[41]

Surendranath Bannerjea attended the Imperial Press Conference of 1909 in London and used it as a platform to express nationalist views. The press claimed that the 'only man who has reason to regret Bannerjea's visit to England is Baron Reuter ... The Press Conference was generally said to be a fiasco and Mr Bannerjea was the only one of the press delegates to have caught the popular imagination.'[42] He also gave a popular lecture in Caxton Hall over which Sir Charles Dilke presided. However, the tour was marred because, shortly afterwards, Madan Lal Dhingra assassinated Sir William Curzon-Wylie, a civil servant supervising Indian students in Britain.[43] It was alleged that 'it is an open secret that one of the duties of Sir William Curzon-Wyllie at the India office has been to preside over an organisation of police detectives and others who have been watching over the movements of Indian students in Britain ... Curzon-Wyllie has been actively engaged in marshalling evidence against the law student Savarker.'[44] Curzon-Wyllie was one of the members of the Telegraph Committee of 1907–8, and the Savarker brothers subsequently became revolutionary heroes, one of them writing the controversial book on 1857: *India's War of Independence*.

The information panic

A conspiracy was expected and discovered in 1908. The Sepoy Mutiny and the uprisings of 1857 had a profound impact on indigenous society in north India and the future of the British state in India. The core event was the revolt by Indian soldiers of the Bengal Presidency Army that gave the signal for quite wide-scale uprisings throughout northern and central India. The unreliability of the Indian agent was indelibly etched in the British imagination as official narratives, memoirs, and accounts by journalists were produced on a large scale from almost the very onset of the uprisings of 1857. It could well be argued that after the 'Black Hole' of Calcutta few events gripped the British imagination as did the events of 1857. This background of terror, treachery and rebellion among the Indian troops and servants (after sepoys the largest number executed belong to this class in the lists of executions following the uprising starting from 1858) is important because this is a narrative of events in north India on the eve of 1907, the 50th anniversary of 1857.

Lord Morley was appointed Secretary of State for India under a newly elected Liberal government in Britain. Although better known as the biographer of Gladstone and Cobden, it is his biography of Jean Jacques Rousseau that is revealing.[45] What comes across in this work is the basic distrust with which he saw Rousseau's life and work. The introduction and biography is pervaded by a deep fear of libertarianism, of social flux, of the masses and of disorder, fears which may have stemmed from Morley's experience of Ireland and its revolutionary activities in London. Between 1905 and 1907 governments across Europe, Russia and America were becoming paranoid about armed anarchism and militant workers' movements. Russia had the 1905 uprising, followed by a series of anarchist assassinations. The Liberal government in Britain, it could well be argued, would agree to an increase in the powers of the colonial government in India only if it was convinced that libertarian and anarchic disorder was threatening the government in India. What Morley and the Liberal government might not have reacted positively to, nor managed to convince the British public about, would be the fact of rebellion in many sections of the government and people in India, something similar to 1857.

The British Government in India could not afford another rebellion. Indeed such an event did not happen in the sense of a revolt among the Indian troops. What happened was the circulation of rumours regarding a possible second uprising of 1857. The paranoia of government officials was clearly revealed in an incident that received wide publicity at the time. District officials in Bihar began to notice strange man-made markings on trees. Just as before, in the uprisings of 1857, chapatis were circulating from village to village in much of the areas subsequently affected by the rebellion. Officials in 1907 drew parallels, and sadhus and mystics who wander across India came under surveillance and suspicion. 'One of the recent scares was the mud marking of mango trees in Bihar ... the most terrible forebodings were circulated and believed everywhere. Was there to be another Mutiny and agrarian uprising? Hundreds of Sanyasis were arrested'[46] and army cantonments were put on alert.[47] The absurdity of the panic in retrospect did not detract from its immediacy.

Information panics can be mapped in their socio-political dimension through their moments of expression. Sadhus, sanyasis and fakirs were rounded up and arrested. It was charged, as in 1857, that these groups moved across India with impunity and preached revolution to the superstitious masses. An editorial warned that 'Sanyasis, the itinerant yellow robed beggars who have no anxieties have always been suspected whenever there has been a widespread movement of unrest ... if every Sanyasi sees that he is suspected of being a conspirator he may be forced to be so in grim earnest.'[48] Contrary to the representation of the sanyasi and fakir as peripheral and elusive figures, they were central as sources of Indian information and regularly employed as such by the government and army. 'The religious orders and mendicants with absolute freedom to go where they will ... with fellow initiates in every town their powers of underground propaganda may be very great, and their power of evil serious ... a small number of them are undoubtedly concerned in sedition and the murder cult. ... Equally potent are these orders as a machinery for Government secret service.'[49] A number of revolutionaries also escaped police surveillance in this period disguised as wandering mystics. Thus this class of people were as much a British official stereotype of wandering fanatics, as they were a class that performed a number of specific functions pertaining to information and secrecy in contemporary Indian politics.

Nervousness in society

Newspapers reported that there was '... nervousness everywhere among the Anglo-Indian population and the ladies were buying revolvers.'[50] Two Bengalis in western clothes were refused entry into the Allahabad fort because they were thought to be Japanese spies.[51] All the members of the Arya Samaj in Patiala were arrested overnight, probably on the basis of a central list, and charged with sedition.[52] The *Rahbar* wrote that there was a panic created in a European club in Bihar by a soda bottle bursting accidentally.[53] There were several leading officials including the Lieutenant Governors Hare and Baker who suffered 'from what appeared to be a somewhat acute form of hysteria'.[54] Minto writing to Chirol on his heartfelt relief at Sir Bamphylde Fuller's resignation mentions that Fuller was 'completely unsuited for a position full of risk' and that the man was 'hysterical'.[55] A commentator notes in passing that Barindra Ghose was congenitally unreliable and the explanation behind the revolutionary movement lay in the madness of Aurobinda and Barindra's mother.[56] There is much discussion about the levels of stress and morale among the subordinate police[57] and claims that they were over-worked and exhausted.[58] It was alleged by the Inspector General of police that many of the high-ranking Hindu police officers were in league with the agitators.[59] The *Charumihir* wrote of the circularity of cause and effect, 'We have seen much sign of unrest where punitive police have been posted [for example, the Ghurkha regiment in Mymensingh] ... just as Government, by sending punitive police everywhere, has cleared the way for oppressing the people, so a current of unrest is followed over the whole land'.[60] Punitive postings were police troops maintained at the expense of a village.

A contemporary newspaper editorial commented under the heading of 'Mental Phase',

> The matter is really a psychological rather than a military one ... same as that of a composite photograph where one portrait overlays the other, and the first one, then the other leaps into prominence according to slight changes in the focusing of the eye ... so soon as the black overlap can be removed, the truer more charitable view can be uninterrupted.[61]

Others wrote, 'If the anarchist is abroad, so also it would seem the alarmist whose aim is to unnerve Anglo-Indian and the British opinion as to make it talk as though India was on the brink of a dire revolt.'[62] Dignan speaks of 'over-emphasis' in the Rowlatt Committee report in order to justify the repressive measures and of 'not carefully discriminating between acts of common criminals and acts of the revolutionary nationalists.' He also writes of the possibly disproportionate fears aroused by the Indian revolutionaries in the Great War who were forgotten and therefore historically buried as the Allies gained in confidence.[63]

It might well be argued that the panic was as much a product of technological change, determined by the velocity of circulation of information through the falling cost of transmission, as it was 'mental'. Sensationalism and a heightened awareness of the expanding media and the importance of publicity were features of this period. Barindra

Kumar Ghose, one of the leading spirits of the revolutionary extremism in India, showed his belief in the importance of publicity. When the Criminal Intelligence Department raided their location Barindra and other senior members of the group decided to confess. They took the officers on a guided tour of the premises showing them where the bomb-making manuals, ammunition, arms, membership lists and seditious literature were hidden. The importance of publicity for a revolutionary organisation was underlined by Barindra's confession, 'My motive in disclosing these details is principally to place the details of our workshops before the country so that others may follow in our footsteps. I do not think such political murders will bring about our desired regeneration of our motherland. It is a means to educate the people ... There was a wide and persistent demand for one successful political murder to stiffen the back of the people...'[64]

The information order: Informants and information

One of the main reasons for the continuing government panic was recurring information scares and shortages. First, it was relatively starved of information between 1907 and 1912. The sources of information supplied to the government were diverse. Second, there was a conflict between the kind of employee or informant that the government required and the kind of people willing to take up these tasks. Two distinct aspects at the level of the district and province presented problems. First was the problem of accurate information, and archives of newspapers, fingerprints and photographs were built. The second was active surveillance, prevention of crime and an accurate network of informers. The overall problem lay with the collation and processing of these sources of information at an all-India level.

The debate over colonial policing tends to concentrate on comparisons and numbers. Popplewell has argued against the notion of 'Russianisation' of administration in British India in this period. He has shown the considerable competition and conflict between the various branches of the coercive machinery of the state.[65] Arnold agrees that the Indian police was by no means comparable to police regimes in, for example, Nazi Germany or Stalin's Russia. The overlap between province and centre, along with financial constraints, conditioned its existence.[66] Both these authors tend to concentrate more on organisation, numbers, recruitment and budgets. The informant and his information become irrelevant in these histories of police or the intelligence service as administrative institutions. The kind of informer and the content of information are ignored and these often crucially determined the structure of policing and state reactions. The diverse, and in this period, irregularly employed sources of information that fed the state and the media brought them both closer and made them more vulnerable to sensational panics.

The government reconstituted the Thagi and Dakaiti Department to form the Criminal Intelligence Department in 1904, partly in anticipation of growing political trouble. The Thagi and Dakaiti Department was formed in the 1830s to deal with organised crime and the people classified as the Criminal Tribes. A Special Branch had been set up to look after political crime in 1887.[67] In 1901, Irwin, Superintendent of the Thagi and Dakaiti Department in his request for a 'secret service for Government of India, … asked for Rs. 50,000 and a small establishment of carefully selected detective agents, to be used for difficult forms of crime, having ramifications that extend beyond the limits of a single province, as well as for investigating political intrigues.'[68] Letters were sent out to the provincial governments and administrations asking for their opinions.[69] Their replies were considered in the report of the police commission set up by Curzon in 1902.[70] By 1907 there was a strong need felt within the government to form a political branch dealing exclusively with political crime while the Criminal Intelligence Department would continue to look after criminal activities. The overlapping centres of command with their vague boundaries contributed to the confusion in this period. The Director of Criminal Intelligence's attempt to promote an all-India secret service was resisted by the local governments and sections of the bureaucracy.

The Director of Criminal Intelligence argued for a central force and provincial force. He pointed out that the chief centres of the

> Indian Political Movement are Calcutta, Lahore, Poona, New York, London, Paris and perhaps Japan … the necessity for secret agents in London and America has recently been brought to the notice in letters from Dublin and London. The Punjab sends money to Calcutta, which is probably distributed to local agitators in East Bengal and Assam. An outbreak of disturbances in Bengal is reflected in Madras. Lahore and Rawalpindi in the Punjab send money to Peshawar in North West Frontier Province that is again used for stirring up the frontier tribes over the border. Political sadhus or missionaries tour all over India, New York and Paris; send out letters which are used for spreading disaffection in the Army and the civil population; and Shyamji Krishnavarma from London offers prizes and other attractions to those who will devote themselves to preaching the subversion of our rule in

India .... I have employed one or two secret agents to furnish me with direct information about what is going on under the surface, and one of them, the agent at Lahore, has kept us supplied with intelligence that is of real value.

It was a forceful argument. The ordinary police were of little value in this kind of work. The Director of Criminal Intelligence wanted a larger establishment with its nucleus permanently employed: '[T]hese men must be paid from the secret service fund, for if they were brought on a regular establishment their identity would be disclosed at once and their usefulness destroyed.' He explained that it was dangerous to employ temporary men, 'since it is to their advantage to report the continuance of disaffection after it ceases, and thus to retain their pay which will stop when the situation improves.'[71]

However, his request for an all-India secret service was temporarily turned down by Viceregal order. The senior bureaucrats like H. H. Risley and H. Adamson along with Minto noted they were '... not at all convinced of the expediency of spreading throughout India a body of secret police working under the immediate orders of the Director, Criminal Intelligence and unknown to the Local Governments ... little result from the work as regards sedition of the secret agents already employed in a limited degree by the Director, Criminal Intelligence. There is no check on them or their work ... they merely submit sensational reports very little of which can be believed, none of which can be tested, and which never end even in a prosecution. Let it rest for the time.'[72] This was a desperate attempt to contain the emerging panic in January 1908. By May 1908, the secret service was a reality: 'the idea in 1908 was that it [Criminal Intelligence Department] would collate information from all the provinces, and would keep the Local Governments informed. The operations of the sedition-mongers are far more widespread, far better organised, and far more advanced than those of any professional criminals...The range of their activities includes England, America, Egypt and Turkey, and they have no hesitation in allying themselves with our enemies and rivals in any part of the world ... [We] need an all-India service beyond local services.'[73] The problems of information and centralised collation were stark.

The cutting edge of repression

Sedition was the key word of the period. The Indian aristocracy had never been a completely reliable ally to the government. However, they were vocal in their support to the government in opposing sedition. If real absolutism and the lineaments of a repressive regime are sought, the Native States provide the best example. The Nizam of Hyderabad wrote to the viceroy in 1909:

> [T]he matter [*swadeshi*] is one in which the interests of the Government of India and the Indian Princes are identical ... I have abstained from causing alarm by issuing manifestos warning my people against sedition. But a very strict watch has been kept over local officials [and] ... orders have been issued to the police and the district magistrates not to allow any meetings in which there was any likelihood of inflammatory speeches being made. Petty officials and other persons having a tendency to sympathise with the movement have been warned, and some transferred in order to break up any attempt to form a clique. The head of the Educational Department has been specially directed to exercise strict supervision over teachers and students and to prevent their participation in any political movement whatsoever. The Nizam fully believes in prompt extradition of outsiders and the application of Regulation III of 1818 [extradition and deportation of offenders].[74]

The Begum of Bhopal wrote to suggest that every ruling chief to establish or increase the strength of his or her secret police. She added with insight that the 'question arises whether it is better to allow views on the movement to be given a free or controlled vent to them and thus get an idea of the movement, or to suppress the publicity of the views totally and allow a chance to further secret societies.'[75] Baroda wrote to suggest the formation of an all-India organisation to coordinate counter propaganda by the government and reported that since January 1908, some 30 visitors to the state were under strict surveillance.[76] Dewas also wrote.[77] The Maharaja of Bikaner called for the establishment of a Press Cuttings Agency to collate articles on government published all over India.[78] The Maharaja of Jaipur circulated a list of 45 newspapers considered to be seditious and banned in Rajasthan.[79] In contrast, the Maharaja and Maharani of Baroda, were kept under surveillance for meeting with Madame Bikaji Cama, Krishnaji Shyamvarma and other revolutionaries in Paris. The British suspected the rulers of Baroda, Bhopal, and, earlier, the Maharaja of the Punjab, as suspect and antagonistic to their interests. They were suspected of helping finance revolutionary activities beyond the frontiers of India.[80]

Initially a host of not very reliable informants were employed. The government had already been using a number of unconventional informants. Like sadhus, madams and pimps were also used as sources of information. More traditional channels of information were in use, but perhaps, in fundamentally new ways. The world of the bazaar and the prostitute quarters and districts figure prominently in records of the period. Buyers and sellers of various goods flocked to these volatile places where information was widely circulated. The marketplace and the whorehouse were not only linked by this world of rumour but were often contiguous in their occupation of urban space in India. The state employed or paid on the basis of information, pimps, madams and prostitutes. This was a world on the limits of law and through it moved the criminal, soldier, revolutionary and police detective in disguise. The revolutionaries also met in the discreet quarters of prostitutes.

There is a description of a meeting of revolutionaries that was successfully raided by Criminal Intelligence Department detectives who used the services of the prostitute next door. Young revolutionaries often took up rooms in the red light district, which was also an information bazaar in which men and news from all over the empire circulated. For example, there was Hamesha Bahar Begum who earned a 'little pocket money in supplying the Indian secret service agent with Afghan inner news. She had the favour of the police and ran a brothel.'[81] In the early part of this period the sources that directly fed information to the state were very diverse. The Maharajadhiraj of Bardhaman, for example, was a source of information regarding the Hindu sabhas emerging by 1912.[82] He had the ears of the Viceroy. Informers seem to be an open category in the official files.

In their searches the police considered material like nationalist songs, poetry and photographs. Poems by Rabindranath Tagore and writings of Aurobindo Ghose were included in the list of proscribed and suspect materials.[83] In their press searches the police usually went for the register of subscribers.[84] Photograph studios had mushroomed and regularly advertised their services as well as photographs of Dadabhai Naoroji, Balgangadhar Tilak, Maharaja of Baroda, Justice Chandramadhav Ghosh, Rashvihari Ghosh, Aurobindo Ghosh, Lala Lajpat Rai, Bepin Chandra Pal, Liakat Hussain and other nationalist leaders.[85] An average search would involve the following finds: private letters, nationalist songs, a diary of a student dating from 1905 in which the author says he took active part in the *sankirtan* of 30 Ashwin, a copy of the *Karmayogin*, and photographs of leaders.[86] Often photographs were the single most incriminating material discovered. In Bombay, photographs of leaders were banned from exhibition and distribution.[87] Group pictures and college photographs were often extremely useful to the police.[88]

From Informant to Informer: The disappearing face of the 'native informant'

Everybody was a potential informant, but the professional informer was a dubious and thoroughly unreliable dealer in information. They rarely figured on the official payroll and most of them never made the official or permanent grade. Any study based on employment rolls or strictly institutional figures would ignore the realities and vulnerabilities of information gathering and collating. The police employed these informants to penetrate various political and revolutionary groups and activities, especially after 1905. At the opposite end of the spectrum was the aristocracy. They were sometimes recruited from within secret revolutionary organisations. They were spectacularly important figures in the events that were reported in the daily press and in the fortnightly correspondence between the Viceroy and the Secretary of State for India. Huge conspiracy cases sometimes involving as many as 100 accused rested on accounts built up by these informers. While the Special Tribunal Acts of 1907 combined with a rigorous application of Sections 124 A and 153 of the Indian Penal Code had ensured that special tribunals appointed by the High Court would be the final court of appeal, confessions extracted in police custody increasingly got reported with outrage in the media with allegations of torture and the judiciary rarely punished on the basis of confessions.

Professional Informers sometimes withdrew their testimony in court and whole gangs were charged by the police when it was found that a series of police raids, for example in Allahabad and Benares, were based on false information systematically fed to the police by professional gangs of misinformants. Casual informers were going to prove increasingly unreliable. In Allahabad in 1908, hundreds of houses were raided in the middle of the night over five days. These were the homes of professionals, prominent and respectable citizens such as legal pleaders, lawyers, editors, doctors and so on. It soon emerged that there was a gang of misinformers operating in the city who systematically fed the Criminal Intelligence Department and the police false information that finally led to these disastrous house-to-house searches.[89] In Lucknow, one Jogendranath Mazumdar, on whose information a number of house searches were conducted, gave a statement before the magistrate stating that he belonged to a 'body of men whose business it was to harass the police', that is, professional misinformers.[90]

The Revolutionaries and the Criminal Intelligence Department often seemed very similar in their tactics. The *bhadralok* dacoit and the

Criminal Intelligence Department officer raided houses in the middle of the night, used torture to extract what they wanted and both employed informers. Criminal Intelligence Department officers could be shot inside the High Court.[91] Newspapers warned, 'The shadowing of suspects is in the ordinary routine of police work ... there is shade within the shadow, that is, spies themselves are spied upon by a set of informers of a deeper dye.'[92] Influencing witnesses and manufacturing of evidence was another issue. In the Midnapur Enquiry set up after the case collapsed, the officers charged with improper conduct were found to be trying to coerce a key witness, a young boy called Bonomali Das.[93]

Lord Hardinge objected strongly to 'indiscriminate house searches on insufficient information. The Government of Bengal should be censured by the Government of India ... these searches fairly exasperate the population.'[94] The government and more immediately the judiciary reflected a growing distrust of these sources and the public panic they could easily generate. The state increasingly preferred its own intelligence employees. However, the identity and even, residential address of officers in the Criminal Intelligence Department was sometimes published in the media in the early part of the period. Thus, the official informer Ashutosh Biswas was attacked at his residence during the Howrah Conspiracy trial.

As a contemporary newspaper put it, the shadows were themselves shadowed. An underworld of ruthlessly professional, interchangeable informers and committed cadres came to light. The local press often reflected snippets of strategic information about these various opponents. The anatomy of violence discussed earlier can be further investigated.[95] There was often a common pool of information circulated via the media. Private telegrams from the Viceroy to the Secretary of State containing extracts from letters sent by the Special Committee on 'separate representation' of the Muslim League were leaked to the press.[96] Information regarding the names of investigating officers, informers and approvers and was regularly published. Equally regular were reports in the press about nationalist meetings and events. The commonality of information is unsurprising if one can see the community of the actors involved. In a lot of cases the revolutionary was an employee or the son of an employee of the state.[97] The informer turned out at times to have been in his employ or dependent on him.[98]

The panic in government in 1907–8 may have resulted in indiscriminate policing that made this aspect of the state more visible. Thus to contemporaries, the British State in India suddenly appeared as a state that had overdeveloped in the area of policing, repression and surveillance. Parallels were drawn in the press with the Czarist police state in

Russia. The Indian informant that Bayly has outlined for the eighteenth century was reduced through this process into a professional informer. In a speech to the Royal Commonwealth Society in 1936, the notorious ex-Commissioner of Police in Calcutta, Charles Tegart, criticised the popular belief that the police went about raiding houses and arresting suspects based on unreliable and temporary informers. The informer employed by the police and the Criminal Intelligence Department, he argued, was a special operative trained to infiltrate revolutionary groups and was employed over a long period of time.[99] This was not the case until after 1910. As late as 1912, six months before the bomb thrown at Lord and Lady Hardinge, the Inspector General of Police in Bengal wrote confidentially to the Chief Secretary, 'there can be no doubt whatsoever ... the nucleus of the department [Intelligence Bureau] must be a permanency.'[100] The impermanent informer and the increased velocity of information circulation determined the contours of the panic in 1907–12.

Aetiology of the bomb

Increased police activity and repression in India soon raised questions for the Secretary of State in the British parliament. Sir Henry Cotton tabled questions and wrote a letter to the *West Minister Gazette*:

> [C]an public opinion in England approve a sentence of seven years hard labour for 'dispatching a seditious telegram', or of five years deportation in another case for 'exhibiting and commenting upon seditious photographs in a railway carriage?' In a third case the assistant secretary of a 'swadeshi' steam navigation company was deported for six years. In another case a Muslim editor was sentenced to two years hard labour ... for "publishing a seditious article", the said article consisting of comments on the methods of education in Egypt.[101]

Proscription often led to wider circulation: Ganesh Balwant Modak published a fortnightly called *Swaraj Sopan* in which he published an article called the 'Etiology of the Bomb in Bengal'. This was banned and Modak arrested.[102] Later that year the W. H. Stead, editor of the *Review of Reviews*, republished the article in England. Its copies were seized at Bombay by orders of the Government of India.[103] A reverse example would be the consideration of translations from English books and newspapers to be seditious. Kelkar, editor of the *Mahrattha*, was charged for reproducing an article by W. J. Bryant called 'British rule in India'. It was published in the New York *Sun* on 28 July 1906. Kelkar, in his defence, stated that he reproduced the article along with the *Hindu*, *Madras Standard*, *Bengalee*, and the *Amrita Bazar Patrika*.[104] Similarly, a book alarmingly titled *Ghadr* or mutiny later turned out to be a translation of one of the standard works on 1857.[105]

The suppression of presses proved quite difficult. The banned *Yugantar* newspaper disappeared from Bengal to reappear in Lahore in December 1909. It was typewritten or printed on a postcard. It contained the shortest recipe to make a picric acid bomb, prescribed bombs as the best means of co-operation with the government and wished all its readers a very happy New Year. It was posted at the Dabbi Bazar Post Office at Lahore.[106] Pamphlets kept appearing in various places including Ireland, Berlin and Paris. *Talwar*, printed in Berlin, reached Lahore through the English mail. It had Madan Lal Dhingra's photograph on the cover. He was the assassin of Sir William Curzon-Wylie and his companion Dr Lalcaca photograph on the cover. It was immediately banned.[107] Similarly, pamphlets preaching violence and extolling Dhingra's crime reached newspapers and individuals in India and England from Paris.[108] The Orange Society in Ireland, allegedly, also circulated pamphlets extolling Dhingra.[109] Shyamji Krishnavarma, an extremist leader, published the banned *Indian Sociologist* from England. He later shifted to Paris and in case of suppression of his paper, changed the printer in England. Similarly, the *Justice*, published in England, was banned in India. There were questions in the House of Commons as to why it was banned in India and also how it reached India by mail and was duly 'circulated, though they bore the name of the paper on the wrapper in flaming red letters an inch long.'[110]

There is little justification for calling these various channels of communication traditional or modern. Some were older than others but not necessarily any less technologically aware or progressive in terms of communication. The problem lies with the fact that the written and the oral, the sign and the symbol had become crucially enmeshed with print, publishing and the camera. These traditions borrowed from each other extensively. Pamphlets were printed in relatively large centres and distributed in the countryside, consciously using popular and old tunes as well as poetical images. In contrast, many of the judgements and the arguments used by the prosecution in cases of sedition involved elaborate translation and investigation of literary and social metaphors and styles. Indeed, the 'information panics' generated among different branches of the administration in 1907–8 because of a fear of a second mutiny led the government into nearly impossible areas of definition and translation of sedition. In one case the government prosecutor declared a verse describing Morley as a cunning jackal (*dhurtya srigal*) to be seditious. The defence held that it was a traditional Indian compliment. Photographs of nationalist leaders were declared to be seditious, as were some items of clothing, and even pornography. The very fact that parliament in London questioned Morley about the harshness of the conviction of the author of the verse mentioned above shows how sensitive the Liberal government in Britain had become to charges of Russianisation.

Most prominent among the cases that collapsed was the Midnapur Conspiracy Case. In late June 1909, the High Court acquitted all the accused and passed strictures on the 'methods adopted by the police and the officers responsible for prosecution'. An enquiry was set up. Aurobindo Ghose's speech at Beadon Square, Calcutta, stressed the issues involved that included misinformation, misrepresentation of facts, use of torture to extract confessions, recruit approvers and influence witnesses. He said

> no subject is safe in this country ... if they [Government of India] were informed by the police (who have distinguished themselves at Midnapore) or by information as tainted – the perjurers, forgers, informers and approvers – that such and such men have been seditious, or becoming seditious, or might be seditious or that their presence in their homes was dangerous to the peace of mind of the CID, against such information there was no safety in this country ...'[111]

At the Jhalakati Conference of 1909 Aurobindo Ghose reiterated his views, claiming that the Midnapur case was 'the standing and conclusive proof of the value of police information. ... We have had charges of sedition, dacoity and violence brought against the youth of our country: charges such as those which have seen breaking down and vanishing into nothing when tested by a high and impartial tribune'.[112] The *Amrita Bazar Patrika* echoed him and asked, 'were not all forces of an irresistible bureaucracy apparently harnessed for the purpose of crushing the gentry and aristocracy of Midnapore?'[113] The informers in the case were dismissed by the High Court as unreliable. One, Rakhal Chander Laha, was described as a 'drunkard' while the other, Abdur Rahman, was called a 'butcher' and a 'peddler' (hawker). The Court called Rakhal a 'pest of society' and Abdur was 'a man of no social position'. It was on their information that the authorities were convinced that 'Raja to beggar in Midnapore were manufacturing bombs to murder' the British authorities. The government was not happy.

Abdur was recruited from a *lathiyal* society in December 1907 at Rs. 25 a month. That was a large sum of money for a man in his position and it is not surprising that that he manufactured evidence and information to suit the fears of his employers. Rakhal was a different case. He claimed in his defence before the court that he was 'lying drunk, was brought in, and was pummelled into compliance.'[114] Rakhal confessed in court that the information he had given in the conspiracy case was false. In support of his statement he pointed out that underneath every paper containing his written information could be found words in his own handwriting, 'What I state here is false!'[115] He was sentenced to rigorous imprisonment for five years and a fine of Rs. 3000.[116] This was not a special feature of the Midnapur conspiracy case.[117] A number of high profile conspiracy and dacoity cases instituted by the Criminal Intelligence Department fell apart under scrutiny. The Alipur Bomb Conspiracy Case fizzled out. Though Barindra Ghose was sentenced, his brother Aurobindo was acquitted. The main approver Noren Gossain was murdered in Alipur Central Jail.

The number of cases that collapsed multiplied over 1909–10. The Salimpur case ended with the acquittal of almost all the accused towards the end of May 1909.[118] Similarly, in the Barrah dacoity case, the court remarked that 'grossly improper influences have been at work and there has been deplorable interference with the evidence'. The Natore Mail Robbery Case and the Mymensingh Conspiracy Case showed similar features.[119] The Rawalpindi Riots, Trivandrum Riots and the Bighati dacoity were also cases that fizzled out.[120] The Haludbari dacoity case promised to fizzle out, as the approver Shailendranath Dass proved difficult, charged that his confession was made under duress and his conditional pardon was withdrawn.[121] The *Advocate of India* commented on the conspiracy theory imposed on the Ahmedabad bomb case, '...there is nothing to connect it with the chain of recent crimes which disturbed the other side of India and not a little evidence to suggest that it is not an outcome of any organized conspiracy of any kind ... we hope ... no handle will be given to scaremongers who are always with us to turn ... the country upside down and fabricate conspiracies and plots which exist only in their heated and fantastic imaginations.'[122] The *Amrita Bazar* gives a counter-typology: the 'inevitable' bomb; the confession of the accused youth under police custody; the alleged threatening of the Magistrate's life; an irresponsible inspector or Criminal Intelligence Department agent and the unreliable informer.[123]

Information and control of information

In 1911, Governor General Lord Hardinge complained of the lack of independent information that had traditionally come from petitions made to the government.[124] As the boycott of government developed, these traditional sources of information dried up. The Government in India introduced the Seditious Newspapers Act in 1907. It was aimed at suppressing a whole class of inexpensive revolutionary newspapers. A series of very high profile public trials were initiated against newspapers like *Yugantar*. Huge conspiracy cases were also started against sometimes as many as 1500 revolutionary suspects. However, by 1911 officials were admitting that these huge trials were colossal *tamashas* and failures. High-ranking police officials remarked morosely 'we cannot very well give up the attempt to govern because of the Chief Justice.'[125] It was admitted that not only did the accused receive a lot of publicity, and those that were deported became martyrs, but the biggest flaw in the system was that the circulation of seditious material was not checked. In 1910 a new press act was formulated and officials noted with satisfaction that it had affected the entire class of seditious newspapers. The simple formula applied was to make editors of newspapers deposit a large sum with the government. This sum would be forfeited if the newspaper were found to have published anything remotely seditious. It was noted with satisfaction that revolutionary newspapers such as *Karmayogin*, *Bandemataram*, *Yugantar* and *Kesari* had folded up within six months of the passing of the act. The government waited from 1907 till 1910 to implement a strategy to curb the press. Yet this method of control had been practised against the Anglo-Indian press as far back as before 1835. One of the reasons for this delay was the importance of these newspapers as sources of information.

The bi-weekly newspaper *Karmayogin* reveals the strategies of a popular newspaper of the new generation. *Karmayogin* circulated from 1906 until its suppression around 1910: Its printing press was confiscated, the publisher and printer imprisoned for sedition. It began as a channel for Aurobindo Ghose's writings after the Alipore case. Ghose's lectures were dialogues between eastern and western notions of nation, patriotism, sacrifice and government: an internationally aware rhetoric. Aurobindo also helped in translating Sanskrit and Bengali literature and the paper was strident in its brand of *bhakti* religiosity, combining sacrifice and devotion to the nation with salvation and dharma. It advertised nationalist slogans, photographs and cures. It regularly reported on police activities and reported speeches of prominent nationalist leaders and advertised the venues of nationalist meetings. The Viceroy noted that the *Karmayogin* provided useful information about the state of revolutionary thought and activity and this was the reason why it was permitted to publish as long as it did. In short, *Karmayogin* was a purveyor of information in a situation of high demand and its situation was similar to other newspapers reprimanded or suppressed for sedition. Most newspapers circulated extracts from other local as well as international newspapers, W. H. Stead's *Review of Reviews* was a very popular digest of international news and information for radical movements and events around the world.[126] In 1911, Hardinge admitted that if these newspapers had not been important as purveyors of information then the government would have shut them down much earlier. Along with the need for a steady and trained base of informers and reliable information was the Government of India's growing need to coordinate between the various provincial governments. An all-India news service along the lines of Reuters was called for that would bring together various happenings all over India. In 1910, an organisation was established to provide a centralised news service to senior officials all over India.

The second conflict along with the need for right information was the British State in India's need for the right informant and the right employee. The difficulty with employees emerged very clearly in the Telegraph Committee's report of 1907. It called for the adoption of a policy that would discriminate against men from Bengal and Maharashtra. The reason given was that they were not physically equipped to deal with the work of the telegraphs. All the orphanage and Anglo-Indian schools were written to asking for young recruits who should receive a preliminary training while still in school. Most of the orphanages and the lower class of missionary schools already ran telegraph-training classes. The more upmarket schools, however, pointed out few of their pupils wished to join the post and telegraph service because their initial pay as well as job prospects were much better with the police, the railways or the private sector.

Our analysis suggests that the use of the conspiracy threat allowed the government to expand certain sections of the executive and judiciary. More and more people were inducted into an expanding Criminal Intelligence Department and new branches were added to the main police such as the River Police during 1907–10. The emergence of an exclusive City Police by the beginning of 1912, and an expanding Finger Print Bureau and Passport Department, augmented the main police. Popplewell[127] has argued persuasively against Dignan[128] to show that in terms of strength, finances and numbers, the growth of the police force in India was not very marked up to the start of the First World War. However, what he ignores is the question of publicity and visibility. Starting from 1907 a series of random late night house searches began to be conducted in a wide and indiscriminate fashion. These nocturnal visitations, sometimes unfruitful, were graphically portrayed in the vernacular press. Historians have by and large ignored the numerous informers not on the official payroll of the police and the sudden increase in recruitment in many branches of the police. This was a very flammable world nationally and internationally, with transnational networks operating across the world. Even ignoring strikes of workers in private industries and concentrating on workers strikes in state and semi-state owned communication industries the numbers involved are large. What should also be emphasised is that these figures involved the European, Eurasian, the 'up-country' men, the Bengali clerks, and the lower castes. Indeed it is possible that these traditional means of enumeration are themselves inadequate to quantify the workers involved. Similar strikes can be observed in Lahore, Agra, Bombay, Allahabad, Madras, Calcutta and Rangoon. Awareness of strikes in London, Paris and other parts was high. A possible means could be to use the categories that emerged more clearly over the 1920s within the structure and organisation of the communication services, that is, categories like inspectors, signallers, peons and women workers who were all organised under distinct factions. But again, the problem remains that much of the classification and subsequent groupings were state sponsored. Certain classes, castes and ethnic groups were discouraged and co-opted over the period of study. This is in contrast with the findings of the Rowlatt Committee with its categories of ethnicity, caste and age. It concluded that high caste Hindus and students were the driving force behind the unrest. In the case of numbers, it records that there was 'definite evidence' of involvement of 1038 persons, only 84 were convicted in 39 prosecutions and of these 30 were tried by special tribunals of three judges constituted under the Defence of India Act 1915.[129] By 1911, the government was admitting that the much-publicised Sedition Cases were a failure. As case after case collapsed in the courts, Hardinge instructed all officials to lay off. The Bengal government was officially censured for its mismanagement of the cases. In 1911, Hardinge was also insisting that the unrest was over and that India was at peace. It was important to signal this because of the impending Durbar to celebrate the coronation of George V. In 1912, Hardinge was himself the victim of a bomb attack. The second phase of revolutionary extremism was said to have begun.

The Russo-Japanese and the First World War

In the Russo-Japanese case, the initial fallout was over the rail and telegraph in China: the *London and China Express* reported the 'hostility of the people of Huna province to telegraph construction', and that a landline from Peking to Kiachtka on the Russian frontier was being advanced.[130] The signing of the Russo-Chinese Telegraph Convention in 1892 brought Britain and China into confrontation. The cable companies adopted a sliding scale of charges depending on where the telegram came from, contrary to the original Anglo-Chinese understanding that China would carry all internal traffic to the 'cable ports', and the cable companies would then take over transmission from these ports. In February 1893, R. Gundry, Secretary to the Committee of China Association, wrote to the British Under Secretary for Foreign Affairs, complaining that an agreement had been reached between China and Russia to lower rates on their lines. The Committee wanted the British Government to intervene because that would benefit only China and Russia, and put other operators at a disadvantage.[131] The Chinese Minister Sheng Taotai explained to the British Consul that the sliding scale policy was an attempt to takeover inland traffic, but this was unacceptable to British interests.[132] A telegraph convention was signed between Britain, India, and China in 1894. In 1897, Britain convinced China to renew the Telegraph Convention of 1894 up to 1904; this convention froze telegraph rates between India and China.[133]

Russian aggression via the telegraphs towards Japan's sphere of influence in 1902 was the final straw, when M. Pavloff, '...without permission from Korea extended a telegraph line from Possiet to Kiong-Hung across Tumen river' and 'desired that the Seul Government should recognize the accomplished fact.' The author published his work on the causes of the Russo-Japanese war before its end, and located the igniting spark at the intersection of telegraphy and political influence. The Korean foreign minister, Pak Chesun, was removed when he attempted to dismantle the lines:[134] where the poles extended there extended the visible power and the sphere of imperial influence. The US welcomed the rise of Japan: '...we of America may rightfully bid the dazed Asiatic seek his salvation from the children of the Rising Sun.'[135] Japan and Russia confronted each other in Manchuria, especially over the Trans-Siberian railway. In Europe the threat came, in part, from the proposed Berlin to Baghdad railway scheme that threatened the security of southern Persia and her communication and oil installations, and, ultimately, India.

The issue of routes and communication occupied centre-stage of many of the conflicts and crises after 1860. Even before the start of the First World War, international Telegraph Conferences met to decide on the security of telegraph cables and lines in the event of a war between

European imperial powers.¹³⁶ The Indian army published, in anticipation, a *War Blue Book* that laid down the measures to be taken by each department in the event of war.¹³⁷ Britain debated the formation of a press bureau and the imposition of censorship after 1900.¹³⁸ Telegraph lines and cables were predictably the first casualties of war. The German wireless chain of stations – the 'German chain' – was more developed than that of the British Empire: with so much energy invested in the telegraph empire, it could not quickly switch to other, quicker and more reliable, industries. Marconi's proposal to build an imperial wireless chain was accepted in June 1915, as was the proposal to build stations in England and Egypt, and in Pune, where a 300 kilowatt wireless station was to be built the Government of India.¹³⁹ The Secretary of State communicated to the Viceroy the urgency felt in the cabinet on this front.¹⁴⁰ In contrast, Germany's telegraph links with the US and the Far East were severed immediately after declaration of war by Britain on 4 August 1914.¹⁴¹ However, Germany had one of the world's most powerful wireless stations at Nauen, which had been operational since 1906.¹⁴²

Britain prohibited the use of codes, imposing censorship over telegrams.¹⁴³ The Germans attacked the two weakest points of the cable system controlled by Britain, namely, the Pacific cable station at Fanning Island, and the Indian Ocean cable station at Direction Island, Cocos. The Pacific cable was repaired after less than a month.¹⁴⁴ The German cruiser *Emden* shelled Madras in September, hoping to destroy the cable station and break the link between Penang and Australia. Having failed, they turned to the cable and wireless stations at Keeling-Cocos Islands. Here too, they achieved only partial success, and Cocos station resumed functioning shortly after the attack. Attempts were also made on Singapore.¹⁴⁵ Though serving British interests during the war, the cable companies were accused of profiteering because they charged full rates on all private telegrams, which now had to be sent without codes or undue abbreviation.¹⁴⁶ The Australian navy destroyed the German Imperial Wireless Chain by destroying stations at Samoa, Nauru in the Marshall Islands, and Herbertshihe on New Pommern Island. The station in Cameroon was captured, while Japan destroyed the station at Kiauchau. The last powerful installation of the German chain fell in May 1915 when South African forces captured Windhoek in German West Africa.¹⁴⁷ The mutiny of Indian troops at Singapore was the revolutionary achievement of the Ghadr Party. Mutiny, the spectre of internal and spontaneous rebellion was transformed into an external incitement, primarily diasporic, towards revolt.¹⁴⁸

The overland system, especially the Indo-European Telegraph Department, was severely threatened. While the lines between Karachi and Teheran were under British control, the routes from Teheran via Constantinople and Europe to London were cut in July 1914.[149] Turkish forces, aligning with the Reich, cut the cable between Fao and Bushehr in November 1914 and took the staff prisoner. They were freed through military action and the break was repaired after less than a fortnight. In response, the department, with military support, captured the Turkish offices at Fao, Basra, and Kurna between November 1914 and February 1915. The Indian Post and Telegraph Department sent out staff in March 1915 to supplement the Indo-European Telegraph Department staff unable to meet military demands.[150] German infiltration, whether through northern Persia or in south east Asia followed the contours of indigenous unrest, using, for example, in the Persian and Turkish case, tribal restiveness for their own purposes.[151]

Though the cable telegraphs returned substantial profits during the war, both because of the increase in volume and the ban on encoding, the landlines in contrast were seriously endangered both in terms of returns and of control and maintenance.[152] The Director General of the Indian Post and Telegraph admitted that during August and September 1914, the first two months of war, there was a large decrease 'all round because of slackness of trade, stoppage of telegrams with enemy countries and the suspension of the deferred telegram system from 4th August 1914'; the last being the most preferred class for private telegrams (see Table 8.1).

The large decrease during the normal months, May to July 1914, was due to the opening of the alternative cable between Aden, Colombo and Penang, thus avoiding India. The larger decrease in value than in numbers was because of the reduction of the Indian transit rate between Madras and Bombay frontiers from May 1914 with corresponding decrease in rates for state and deferred telegrams and to a reduction the

Table 8.1 Table showing effects of war on Indian telegraph messages (percentage of decrease)

Classes	May to July 1914		Aug to Sept 1914	
	Number	Value	Number	Value
State	−60	−76	−94	−90
Private	−40	−52	−79	−60
Deferred	−13	−30	−95	−94
Press	−76	−87	−56	−58
Total	−39	−52	−80	−56

rates of press telegrams between countries west and the east of India.[153] The Indian Government finally allowed, in July 1914, the Eastern Company control over the Bombay to Madras line and sanctioned the lowering of internal and transit telegraph rates. The Department stated that they had 'no reason to consider that the decrease recorded in the number and value of foreign trans-Indian telegrams dealt with over the Bombay-Madras section has been due to any deliberate attempt on the part of the companies to confine the use of the line to cheap-rate [that is, subsidised Press and State] traffic.'[154] However, the number and value of foreign telegrams passing through India and India's share in the Joint Purse showed losses. The total from June 1913 to May 1914 was 357,532 messages consisting of 107,978 words of which, 569,209 was the Indian share. In comparison, the total from June 1914 to May 1915 was 186,047 messages consisting of 3,070,303 words, giving India 386,350 words on which revenue accrued, so the decreases were: – 170,485 in numbers; – 34,685 in words; resulting in – 182,859 in India's share.[155] This calculates to a 47 per cent reduction in numbers, 111 per cent in words, and 32 per cent in India's share of the revenue.

The Indo-European Telegraph Department fared worse than the Indian Telegraph Department. It was one of the main channels of communication between Britain and Russia until 1917. In 1914, State telegrams sent from London for Petrograd were transmitted via the Indian Ocean cables to Karachi and then transferred to the Department's lines to Teheran and subsequently to the Siemen's lines for passing to the Russian Crown's system. A cable was laid through the North Sea from London to Russia.[156] A lot of Indo-European Telegraph Department officers and men left to join the South Persia Rifles under Sir Percy Sykes to defend southern Persia and specifically the telegraph from injury. The pressure of work proved too much for the small department and reinforcements were sent from India in March 1915.[157] Seen in this context, the Dardanelle's campaign was directed less against a resurgent Turkey inspired by, among others, Kemal 'Ataturk', and more for the defence of the telegraph to Britain and to stop the 'Germans who roamed Persia and Afghanistan causing a lot of trouble'. Persia claimed neutrality so the Germans took it as an invitation to raid most telegraph stations under Indo-Persian control.[158] Britain landed forces at Bushehr to protect her wireless and telegraph installations but the hinterland was inaccessible. The Indo-European Telegraph Department suffered heavy financial losses and suspended the Joint Purse Agreement because of the damage to the lines caused by war. The Turkish won Kerman, while Germans attacked Shiraz, Yazd, Isfahan, and looted the Bank of Persia, and Britain evacuated personnel from Persia between 1916 and 1917. In the Persian Gulf section, 15 offices were closed and the number of offices open to the public was only seven.[159] Reinforcements arrived for protecting installations, including cable and wireless stations at Bushehr on the Gulf, but the hinterland was closed to Britain for some time. However, with the help of Indian troops, the crisis gripping British south-Persia abated after 1917.[160] Indian troops played an important role in the war, and censorship was imposed in India through the Defence of India Act. The Chief of General Staff requested for measures to 'prevent the leakage of information as to the nature, composition, or dates of sailing of the forces being despatched from India' in August 1914. This was a 'military war measure of necessity, both to safeguard the transports ... and to prevent the dissemination of information which at the time might have had vital consequences for the National Cause'.[161]

Conclusion

The First World War was a communications crisis, a crisis with many dimensions and consequences. There was a crisis of information: either too much or too little, and a continuous anxiety and growing panic about the reliability of information. In previous histories, the first decade of the twentieth century was distilled to a few grand teleological narratives of nationalism, imperialism, and the First World War. There were several internationalist and nationalist visions of India. There were different kinds of imperialist, European, and South East Asian, engagements with Indians, not least at the trenches. The myth of the 'warrior races' of India suddenly appeared extremely brittle as these people found little rational reason except a fuzzy notion of *Pax Britannica* to flounder in the mud and horror of Flanders and Somme. There were many small wars happening throughout the first two decades of the twentieth century, including in Persia. Many generations were devoted to telegraph departments and services, which suddenly appeared to be threatened. This chapter has demonstrated that technology – especially telegraph technology – was an important element of the period towards the end of the nineteenth century, when the world of electronic communications began to spin out of control, beyond the control of particular governments. Communications as an important factor has been ignored in previous histories and this work readdresses these issues.

Figure C.1 Horizontal imagination: The telegraph station at Jask
Source: The Indian Post and Telegraph Magazine, GPO

Conclusion

Before 1857, it was still possible to imagine a future for the private ownership of telegraph lines, for original invention in India, technological substitution, and perhaps, even for the colonising project itself. Chapter 1 of this monograph shows that the result of the Anglicist and Orientalist confrontation over the nature of knowledge and language influenced the diffusion of knowledge of the principles of electricity in indigenous society, and the impact and nature of innovation in India. The parameters of invention and innovation were defined through law and subsidy, and the nature of knowledge production in the colony was transformed between 1830 and 1857. India, in 1857, was an information-rich world. Described as 'gossip and rumour', Indian information systems appeared as inexplicable phenomena to the British, but were tested means of transmitting certain kinds of information within the country. The rebellion in the Native Army occurred in a technological context, and the character and distribution of the uprisings of 1857 demonstrate that telegraphs were a strategic target.

The uprisings of 1857 saved the Indian telegraph from an ignominious death, allowing reconstruction on a less experimental level: the first system had linear routes and non-insulation of lines, both of which proved to be liabilities during the mutiny. The myth of the overland telegraph as saviour in 1857 renewed interest and investment in telegraphy. Paradoxically, this interest was centred on submarine cable telegraphy as the least destructible of systems, given the unchallenged British naval dominance during the period. The events of 1857 were used to justify both the need for an imperial telegraph system, and direct control by Britain over India, but the myth of the success of the landlines justified the concentration on submarine telegraphy as the most secure means of control. On a more human level, O'Shaughnessy's absence during the height of 1857 and the subsequent attacks on his style of management and technological acumen, saw the 'father of the Indian telegraphs' discredited by the 1860s. Reconstruction after 1857 self-consciously and rigidly adhered to western models, instruments and expertise.

The repatriation of technology to Britain after 1860 combined with relatively high internal rates for private telegrams, which restricted Indian access, while the telegraph department was transformed into just another non-experimental budgeted department of government. The telegraph restricted information flow by high rates and it was easier for Indians to communicate under European names and represent themselves in Europe rather than in India. The telegraph underwent militarisation, Europeanisation and metropolitanisation, reflecting what was happening in all branches of the administration after 1857. The control of the India Office during this period was very evident. Between 1835 and 1875 telegraphy impacted on India in the following ways: first, knowledge of the technology was not diffused and usage implied little more than a skill; second, large imports of machinery from Britain meant that spin-off benefits did not go to India.

After 1860 telegraph technology was standardised and signalling practice was strictly controlled. The telegraph had a significant impact on language, raising questions of context and meaning. Issues such as hearing, reading, secrecy, publicity, error, time, and interpretation enmeshed in telegraphy. Saving telegrams, repeating them, the elaborate records of error, institutions and offices established to scrutinise departmental functions – all concentrated on tracking down the source of error in a message. The process of disciplining the signaller and training the public was achieved over time. If secrecy was insisted upon, there had to be basic structures that allowed its maintenance: the transfer of telegraph offices from rented buildings to specially designed buildings reflected this necessity. Telegraph technology could automatically malfunction and involved issues of interpretation and practice that had consequences for everyday life.

Much of the innovation in telegraph signalling instruments came from signallers themselves, and many an eye, a hand, or ear was damaged serving the tapper. Haphazard recruitment of signallers in Britain between 1865 and 1873 ensured that almost 40 per cent of the senior staff would retire between 1900 and 1904, and promotion was slow. This created increasing politicisation within department employees. The English language telegraph system in India allowed European business to dominate Indian merchants after the cotton crash of 1866, just after the end of the Civil War in US and the resumption of us cotton exports to Britain. After 1870 there was a return to Indian manufacturing, and telegraph workshops were given adequate facilities to test, modify and repair instruments and lines. The Director General recorded in 1873 that 'some of the old Hindu mechanics now possess an amount of skill it would be difficult to surpass anywhere.'[1] The quality of knowledge promoted by the workshops remained confined to the level of mechanical skill, while only a handful of Indian recruits were trained in how to signal at the engineering colleges.

Following the end of the East India Company's rule, India was relatively poor in information, and information cartels such as Reuters and Associated Press dominated international news transmission. Although they were highly subsidised by the British government in India, the vernacular press could not easily afford their rates, and instead relied on digests of news and translations from Anglo-Indian newspapers with more resources. The prevailing argument among officials between 1860 and 1880 was that reductions in internal tariff in India, unless massive, would not be justified by the increase in telegraph traffic. Indian rates were high during those 20 years, and Indian and Anglo-Indian businesses became leaders in the formulation and use of codes and the packing of telegrams, that is, condensing many private, commercial and news messages within a single telegram to avoid high transmission charges. Both the department and the cable companies argued in favour of high rates until this policy was abandoned in favour of a very cheap service and uniform. The imperial project of telegrams adopting the ideal of the Penny Post and representations from commercial groups in India concerning the high rates influenced the department to cut rates in parity with the cable network.

The growth of the telephone and wireless as rival systems of communication, alongside pressure from the parliamentary lobby demanding an imperial government cable system, convinced cable companies to lower their rates. Though the rates were still expensive, these reductions meant a greater volume of traffic. The reduction permitted rapid British capital flows to Australia during its economic crisis in the 1880s, and cable companies returned high dividends after 1875. The growth of telegraph cartels was mirrored by news monopolies, and an international, trans-cultural, trans-regional, and trans-national world appeared a possibility after 1880. This facilitation of communication allowed more vigorous and organised nationalism to emerge within and beyond the Indian National Congress. These strands of multinationalism, transnationalism and nationalism debated the idea of empire, nation state, representation and responsibility. It was an international discourse and many of its figures lived abroad and attended international conferences, engaging with socialist, nationalist, anarchist, and liberal thought. Early Indian nationalists addressed themselves to this international arena. In a letter to Morley, Minto wrote, 'There were indications of the existence in many parts of India of a widespread agitation having as its object the overthrow of our Government. We have reason to believe that the agitation received much support from sympathisers in England, America, as well as other foreign countries.'[2]

The next substantial reduction in rates occurred under Curzon's administration (1898–1905) and combined with the reorganisation of the telegraph department. As a consequence of the changes recommended, the telegraph general strike of 1908 paralysed imperial communications. Though indigenous popular protest destroyed telegraph lines, systemic crisis was achieved only through the combination of telegraph employees. A conjuncture between technology and telegraph labour; the human and the machine. The subsequent emergence of community and racial identity in the process of working class unionisation reflected on a smaller scale the partial democracy, the form of political or rights representation, and the communal award system soon to be introduced throughout India. At another level it illustrated the failed potential for labour mobilisation across class and national boundaries in the telegraph industry. A trans-hierarchical and trans-regional and even trans-national potential for class action was lost through the experience of the strike, and the concurrence of the strike with nationalist agitation.

Simultaneously, a more locally informed and regionally focused information order emerged replacing the larger information agencies. The international lowering of telegraph rates and reduction in press rates loosened the stranglehold of Reuters over imperial and Indian communications. Within India, and within telegraphy as a whole, the information boom created by lower rates saw the state adopt measures that provided more publicity than results. Using local idioms and local information, the anti-partition nationalism in Bengal was initially a peaceful movement involving large gatherings, middle-class students, and a passive boycott of foreign goods. The informal information sector took advantage of the situation, feeding the ensuing panic that was reflected on a global scale. It was a looming international technological crisis that was simultaneously a labour and management crisis, and a political crisis including India as a telegraph hub. The Bengal government reacted to the *swadeshi* and boycott movements with repressive measures that sent the movement underground and along revolutionary channels. Between 1907 and 1910, an attempt at direct repression was replaced by harsher but less visible policing. After 1912 the revolutionary movement had important international dimensions. The fact that after the experience of the first hanging, bodies of revolutionaries were not brought out into the public, and tribunals were set up to judge *in camera*, shows the state trying to contain information and unrest. I argue that the changes in global and national telegraph rates were an important factor in the function and articulation of the information order. British repression dispersed the revolutionary leadership and this allowed the rise of the next wave of leaders who seized the opportunity for mass politics after 1919.

At the onset of war, Mr H. Barton, the telegraph unionist, was imprisoned under the Defence of India Act. He was released in December 1917 after being forced to accept a proposal to divide the Telegraph Association into different branches that were separately controlled.[3] The war, further reduction of rates and the truncation of the imperial telegraph system coincided with the rise of a more integrated national business,[4] economy and mass politics in India. This action by the government shows the threat posed by the politicisation of staff and the importance of Indian and Indo-European telegraph systems in connecting Russia, Britain and the south east, and the association of labour activism with politically dangerous activity. The war like the Russo-Japanese war was over spheres of influence and lines of communication. The telegraphs were also strategic investments, leading to British involvement in west and southern Turkey, southern Persia, and cable ports in Africa and Mesopotamia during the war. The all-red cable system was almost complete except across the Atlantic from Canada to Britain, which was under US control. In contrast, the British imperial wireless station chain was incomplete in 1915. Germany incorrectly imagined that wireless propaganda would be sufficient to sustain it but its more complete wireless chain proved difficult to protect. The all-red proposal was abandoned after the war with the perfection of band wireless. The war destroyed the Joint Purse agreement and the Indo-European Telegraph Department was dissolved in 1931.

On a global scale, India and telegraphy provided the means to maintaining communication outposts and influence beyond the frontiers of formal empire. New imperialism, which was an important concept to contemporaries in Britain in the 1880s, is not mentioned in the *Oxford History of the British Empire*.[5] This book has argued for the validity of this term in the context of telegraph imperialism in many parts of the world. India was involved through the telegraphs in southern Persia, southern Arabia, Burma, and south east Asia. Indian resources were invested outside India during the period when Indian telegraph rates most subsidised trans-Indian traffic. After 1860 telegraphy and railroads emerged as crucial elements of imperial strategy and international relations, and India and telegraphy became key launching pads for the extension of informal empire. Extra-metropolitan discovery and modification played an important role in the development of telegraphy. Britain controlled the Suez, the Arabian sea and the Indian Ocean while Russia offered a railway through itself and Europe to reach London in 1910: the tangible symbols of the great game were the railways and telegraphs. They provided an informal empire within which a variety of interests and new wealth, both intellectual and material, emerged.

The telegraph had crucial political dimensions and formed the basis of an intricate and inherently fragile world order based on issues of sovereignty, territory, apportioning of rates and tariffs, who would build and work the telegraph, and how it could be rendered safe from interruption. Telegraphy attracted power and provided outlets for centralising and imperialistic ambition. This book studies the mentality of telegraphy and its effect on governmentality. Telegraphy promised many things that it did not carry out in practice. In political terms and in terms of the laboratory the representation of metropolitan control was erroneous. Contrary to popular belief, the telegraph did not necessarily lead to greater control by Britain over the colonies. Rather, in some instances, it made Britain more a victim of events elsewhere and committed to the decisions of the man-on-the-spot. The telegraph was not a neutral technology, potentially accessible to everybody in practice, but a tool of power and economic control. The large amount of money extracted from the empire through annual subsidies to the submarine cables made them a joint stock enterprise and explains why the British Post Office lobbied alongside the cable companies against the wireless and telephone. The empire was also, in part, the cable company empire and therefore was potentially very vulnerable to interruption and the transfer of ownership. The telegraph rapidly became a British national concern, and was heavily invested in, in terms of ideology. It was the symbol of Britain's world domination and, after the mid-nineteenth century, the telegraph and the electrical sciences in general represented the most prominent 'marvel' of the day.

The conservatism of telegraph cartels made it difficult to respond to a changing technological context. In terms of scientific imagination, two themes stand out: first, the debate over the electrical metaphor of field versus flow; second, the notion of stasis versus the notion of enervation. Europe believed for a long time, even after Clerk Maxwell, in a world of electrical flows. Ideas of 'action at a distance' or properties of the 'ether' and ideas of an 'electrical fluid' were difficult to unify.

Conclusion 217

The problem of whether or not 'electrical fluid' was an actual physical entity with the properties of a fluid further compounded the dilemma. The difference between a wireless and telegraph imagination was illustrated by architectural construction: while telegraph offices were imperial neo-classical with horizontal expansion echoing poles and wires, skyscrapers reflected the radio antenna (please see Figure C.2).

The zealots of the telegraph age did not foresee its replacement by more efficient forms of technology. Pyenson argued for the continuity of research in the exact sciences, and that the imperial context had little impact on pure science.[6] Headrick rejected accounts that accorded ruptural and revolutionary agency to rail, steam and telegraph technology, and suggested that the strategies and objectives of information systems

Exterior of station at Poldu used by Marconi in his transatlantic signaling.

Figure C.2 A different architecture: Marconi's towers
Source: *The Indian Post and Telegraph Magazine*, GPO

were much older projects and had little ruptural effect in terms of newness. This book finds that in this period the practical implementation of electronics dominated research. The problem of electrical dispersal in submarine cables led to Clerk Maxwell's formulation of field theory. The inspiration for research in the exact sciences came from practical problems while the universality of technology and science was repeatedly tested in the colony. Headrick's evolutionism denied the ideology and finance invested in telegraphy and how electricity appeared as a revolutionising force to its contemporaries. Though the information strategies might have started earlier, telegraphic practice generated its own set of problems and a different array of responses.

Prakash's study of science and the imagination of India echoed Pyenson in its refusal to engage with issue of technological change, the importance of testing its universality in the colony, the domination of a particular technology in the sphere of science, and technological experience inspiring change in pure science. In ignoring the diversity of experience both Indian and European, metropolitan and peripheral that informed enquiry, innovation, and change he did not observe a contingent technology and a contingent modernity. He visualised telegraphy as one of the 'sinews of power' of the modern state to be acquired by the nation,[7] and described it, like Morus, in Benthamite and Foucauldian terms of discipline.[8] The interface between human and technological practice saw the age of telegraphy visualised as the age of nervous machinery. It was imagined as a body with a brain subject to panics and mental fatigue. Telegraphy attracted power and became the key dimension of governance, governmentality and informal empire. In imagining the telegraphs as a liberal instrument of progress and reform, the Victorians also imagined a particular state. This book shows how technology was contested and changing. It also shows how the disciplining of technology and society was inherently fragile and never complete. Both the fetishised technology and the imagined state enabled by it were in crisis. The interplay between the idea of empire, the nation state, and communication technology led to the early nationalists debating in a universal and international context. Commentators on the telegraph pointed out the retardation of business because messages were conveyed only in English in India. China in contrast controlled her telegraphs, allowed transmission in the vernacular, and registered significant telegraphic and business expansion.[9] As the imagined state and its propaganda of technological success faltered in practice, a widening spectrum of nationalist thought emerged, precisely at moments of greater communications facility.

Notes

Introduction: 'What hath God wrought!'

1. *The Book of Common Prayer and Administration of the Sacraments, and other Rites and Ceremonies of the Church according to the Use of the Church of England together with the Psalter or Psalms of David, Pointed as they are to be Sung or Said in Churches*, Oxford: John Baskett, 1721, [*The Fourth Book of Moses*: Numbers] Balak's Sacrifice/Balaam blesses Israel, Chapter XXII, Verses 23, 24.
2. '[T]he attribute of a specific form of social organization in which information generation, processing and transmission become the fundamental resources of productivity and power.' This is the hegemonic logic of IT time, as the telegraphs were hegemonic from 1850 to 1900; Manuel Castells, *The Information Age: Economy, Society and Culture*, Vol. I: *The Rise of the Network Society*, Cambridge, MA and Oxford: Blackwell, 1996, p. 21.
3. Cf. Tom Standage, *The Victorian Internet: the Remarkable Story of the Telegraph and the Nineteenth Century's Online Pioneers*, London: Weidenfeld & Nicolson, 1998; Brian Winston, *Media Technology and Society: A History from the Telegraph to the Internet*, London and New York: Routledge, 1998.
4. I am very grateful to Dr A. Mein, Westcott House, Jesus Lane, Cambridge University, for his generous help with this question.
5. Kalidas Maitra, *Electric Telegraph ba Taritbartabaha Prakaran*, Srirampur: J. H. Peters, 1855.
6. Kalidas Maitra, *Vaspiyakal o Bharatiya Railway: The Steam Engine and the East India Railway containing a History of India, with a Chronological Table of the Indian Princes, from Judister down to the Present Time with a Description of the Places and their Histories through which the Railway Passes, with a Coloured map and Many Illustrations*, Srirampore: J. H. Peters, 1855.
7. The variety of debates and sources that manufacture what is valid information in the public arena: cf. C. A. Bayly, *Information and Empire: Intelligence Gathering and Social Communication in India 1780–1870*, Cambridge: Cambridge University Press, 1996.
8. C. A. Bayly, 'Informing Empire and Nation: Publicity, Propaganda and the Press, c. 1880–1920', in Hiram Morgan (ed.), *Information, Media and Power through the Ages*, Historical Studies XXII, Dublin: University College Dublin Press, 2001, pp. 179–201.
9. For example, Kishanlal Sridharani, *The Story of the Indian Telegraph* (Preface by Jawaharlal Nehru, the first Prime Minister of India, who described the Telegraph Department as one of the oldest public utilities in the world), Delhi: Government Printing, 1953.
10. I. G. J. Hamilton, *An Outline of Postal History and Practice*, Calcutta, 1910; E. Bennet, *The Post Office and its Story*, London, 1912; G. R. Clarke, *The Story of the Indian Post*, London, 1921; E. Murray, *The Post Office*, London, 1927.

11. Sridharani, *The Story of the Indian Telegraph; The General Post Office Centenary Volume*, Delhi: Government Printing, 1954, including articles by I. K. Gujral, Suniti Kumar Chatterjee and R. C. Majumdar. For a more detailed and explicit critique of the 'legacy' of telegraph communication see, D. K. Lahiri Choudhury, 'O'Shaughnessy and the Telegraph in India, c. 1836–1856', *The Indian Economic and Social History Review*, vol. XXXVII (3), July–September 2000, pp. 331–59.
12. Lewis Pyenson, 'Cultural Imperialism and Exact Sciences Revisited', *Isis*, vol. 84 (I), March 1993, pp. 103–10.
13. Bruce J. Hunt, 'Doing Science in a Global Empire: Cable Telegraphy and Electrical Physics in Victorian Britain', in Bernard Lightman (ed.), *Victorian Science in Context*, London and Chicago: University of Chicago Press, 1997, pp. 312–33.
14. Ibid., p. 320.
15. E. Hobsbawm, *The Age of Empire: 1875–1914*, London: Weidenfeld & Nicolson, 1987.
16. V. I. Lenin, *Imperialism, the Highest Stage of Capitalism*, in *Selected Works*, 3 vols, vol. 1, Moscow: Progress Publishers, 1975.
17. D. K. Fieldhouse, '"Imperialism": An Historical Revision', *The Economic History Review*, Second series, vol. XIV, 1961.
18. E. Stokes, 'Late Nineteenth Century Colonial Expansion and the Attack on the Theory of Economic Imperialism: A Case of Mistaken Identity?' *Historical Journal*, vol. XII, 1969.
19. John Gallagher and Ronald Robinson, 'The Imperialism of Free Trade', *The Economic History Review*, Second series, vol. VI, no. 1, 1953.
20. Bayly, *Information and Empire* pp. 143, 149, 171–3, 316.
21. Andrew Porter (ed.), *The Oxford History of the British Empire: The Nineteenth Century*, vol. 3, Oxford: Oxford University Press, 2001.
22. R. Koebner and H. D. Schmidt, *Imperialism: The Story And Significance of a Political Word, 1840–1960*, Cambridge: Cambridge University Press, 1965.
23. See, cf. C. A. Bayly, *Information and Empire: Intelligence Gathering and Social Communication in India 1780–1870*, Cambridge: Cambridge University Press, 1996, pp. 143, 149, 171–3, 316.
24. Benedict Anderson, *Imagined Communities: An Inquiry into the Origins of Nations*, London: Verso (rev. and extended edn), 1991.
25. Peter Matthias (ed.), *Science and Society, 1600–1900*, Cambridge: Cambridge University Press, 1972; A. E. Musson (ed.), *Science, Technology and Economic Growth in the Eighteenth Century*, London: Metheun, 1972; S. Bhattacharya, 'Cultural and Social Constraints on Technological Innovation and Economic Development: Some Case Studies', *Indian Economic And Social History Review*, 3 (3), September 1966, pp. 240–67.

1 From Laboratory to Museum: The Changing Culture of Science and Experiment in India, c. 1830–56

1. The colonial state is not just the East India Company's government but also its role as a revenue collector, administrator, controlling justice, and formulating policies, which have long-term expected and also unforeseen consequences. So the colonial state is more than the sum of any particular government.

2. S. N. Mukherjee, *Sir William Jones: A Study in Eighteenth Century British Attitudes to India*, Cambridge: Cambridge University Press, 1968, pp. 3–4.
3. S. Sangwan, *Science, Technology and Colonisation: An Indian Experience 1757–1857*, Delhi: Anamika Prakashan, 1991, p. 43.
4. Partha Chatterjee, 'The Disciplines in Colonial Bengal', in *idem* (ed.), *Texts of Power: Emerging Disciplines in Colonial Bengal*, Calcutta: Samya, 1996, p. 12.
5. O. P. Kejariwal, *The Asiatic Society and the Discovery of India's Past 1784–1838*, Delhi: Oxford University Press, 1998, pp. 226–7.
6. Ibid., p. 224.
7. John Drew, *India and the Romantic Imagination*, Delhi: Oxford University Press, 1987, p. 77.
8. Cf. Nair (ed.), *Proceedings of the Asiatic Society 1817–1832*.
9. Home Department, Public Proceedings, Press, 14 April 1834, no. 20. Letter from the editor of the *Journal of the Asiatic Society*, 9 April 1834. National Archives of India, Delhi [henceforth NAI]; *Proceedings of the Asiatic Society*, May 1836. Asiatic Society Library [henceforth ASL.]; Rajendralal Mitra, A. F. R. Hoernle, P. N. Bose, *Centenary Review of the Asiatic Society 1784–1884*, Calcutta: The Asiatic Society, 1885 (repr. 1986) Part I, Appendix B, pp. 51–2.
10. R. S. Rungta, *Rise of Business Corporations in India 1851–1900*, Cambridge: Cambridge University Press, 1970, p. 9.
11. Mitra, Hoernle, Bose, *Centenary Review*, Part I, Appendix B, pp. 71–3.
12. Home Department, Public Proceedings, 28 January 1834, no. 28. From the president of the Medical and Physical Society; Home Department, Public Proceedings, 3 March 1834, nos 29 and 29A. From the Dinapore Branch of the Agricultural and Horticultural Society. All requests for financial support were turned down. NAI.
13. Cf. E. Said, *Culture and Imperialism*, London, New York: Vintage, 1994, p. 130.
14. Deep Kanta Lahiri Choudhury, 'Communication and Empire: The Telegraph in North India c. 1830–1856', unpublished diss. Jawaharlal Nehru University, Centre for Historical Studies, Delhi, 1997, esp. pp. 48–51 with tables and graphs.
15. Bayly, *Information and Empire*, p. 370.
16. Cf. D. Ludden, 'Orientalist Empiricism: Transformations of Colonial Knowledge', in C. A. Breckenridge and P. van der Veer (eds), *Orientalism and the Postcolonial Predicament*, Delhi: Oxford University Press, 1994, pp. 250–78.
17. B. S. Cohn, *Colonialism and its Forms of Knowledge*, Princeton: Princeton University Press, 1996, p. 7.
18. S. Ghose, 'Introduction and Advancement of the Electric Telegraph in India', unpublished Ph.D. diss., Jadavpur University, Faculty of Engineering, Calcutta, 1974, p. 247.
19. *Proceedings of the Asiatic Society*, May 1837. ASL.
20. Home Department, Public Proceedings, Committee of Public Instruction, 1 July 1835, no. 14. NAI.
21. Mitra, Hoernle and Bose, *Centenary Review*, pp. 51–2.
22. Ibid., Appendix, p. 83.
23. Matthew H. Edney, *Mapping an Empire: The Geographical Construction of British India, 1765–1843*, Chicago: Chicago University Press, 1997, p. 262.

24. Cf. on law, R. Rocher, 'British Orientalism in the Eighteenth Century: The Dialectics of Knowledge and Government', in C. A. Breckenridge and P. van der Veer (eds), *Orientalism and the Postcolonial Predicament*, Delhi: Oxford University Press, 1994, pp. 214–49.
25. J. W. Kaye, *The Life and Correspondence of Charles, Lord Metcalfe*, (new and revised edn) vol. II, London: Smith, Elder and Co., 1858, p. 147.
26. Cf. Jurgen Habermas, *The Structural Transformation of the Public Sphere*, (trans.) Thomas Burger with the assistance of Frederick Lawrence, Cambridge: Polity, 1989 (first pub.1962).
27. *Dictionary of National Biography*, London, 1895, vol. 42, p. 310.
28. *Calcutta Annual Directory*, Calcutta, 1834. National Library [henceforth NL].
29. Cf. C. M. Shepard, 'Philosophy and Science in the Arts Curriculum of the Scottish Universities in the Seventeenth Century', unpublished Ph.D. diss., Edinburgh University, 1975.
30. W. B. O'Shaughnessy, 'Memorandum Relative to Experiments on the Communication of Telegraph Signals by Induced Electricity', *JASB*, vol. JLVIII, September 1839, pp. 714–31. ASL.
31. Cf. *JL*VIII, 147; *JL*VIII, 714; *JL*VIII, 732, 838. ASL.
32. J. Falconer, 'Photography in Nineteenth Century India', in C. A. Bayly (ed.), *The Raj: India and the British 1600–1947*, London: National Portrait Gallery Publications, 1991, pp. 264–77.
33. *Society of Telegraph Engineers*, proceedings volume, 1876, p. 12. General Post Office Collection, Calcutta [Henceforth GPO].
34. *Journal of the Asiatic Society: JL*III, 145; *JL*VIII, 147; *JL*VIII, 714; *JL*VIII, 732, 838; *JL*VIII, 351; *JL*IX, 277; *JL*X, 6; *JL*XII, 1066; *JL*XVI, 177, 557. ASL.
35. Mitra, Hoernle, Bose, *Centenary Review*, Part I, Appendix B, pp. 53, 87–9; also *Proceedings of the Asiatic Society*, May 1839, 1842 and 1846. ASL.
36. Home Department, Public Proceedings, Committee of Public Instruction, 5 August 1835, no. 15, from the Secretary with enclosures, 28 July 1835. NAI.
37. Home Department, Public Proceedings, Committee of Public Instruction, 30 June 1835, no. 6; also 8 July 1835, nos 4 and 5, from the Secretary, the General Committee of Public Instruction. NAI.
38. Home Department, Public Proceedings, 4 April 1850, no. 49. From Lieutenant Colonel W. N. Forbes, Mint Master and Superintendent of Government Machinery, to Captain Scott, Secretary to the Military Board, 19 February 1850. NAI.
39. Home Department, Public Proceedings, no. 4, 11 April 1850, no. 429. From Sir Henry Elliott, Secretary, Government of India, endorsing recommendations of the Military Board. NAI.
40. Home Department, Public Proceedings, 23 April 1852, nos 12 and 13. From the Secretary, Government of Bengal, to the Secretary, Government of India, with enclosures including O'Shaughnessy's 'Abstract account of the construction of the experimental telegraph line from Calcutta to Kedgeree', and a Minute by the Governor General. NAI.
41. Home Department, Public Proceedings, 23 April 1852, no. 13, from O'Shaughnessy, to J. P. Grant, Secretary, Government of Bombay, 10 February 1852. NAI.
42. Home Department, Public Proceedings, 21 June 1850, no. 28. From the Military Board to Major General Sir J. H. Littler, with enclosures, 4 June 1850. NAI.

43. Home Department, Public Proceedings, 23 April 1852, no. 13. From O'Shaughnessy to J. P. Grant, Secretary, Government of Bombay, 3 March 1852. NAI.
44. Also called the single needle galvanoscope. It was used together with Cooke and Wheatstone's 'ABC' instruments. A battery and a reversing handle, or two tapper keys, the motions to the right and left end of the index corresponding to the dashes and dots of the Morse alphabet worked it. The needle was of soft iron and was kept magnetised by the action of two permanent magnets. It was widely used in England in the 1850s.
45. The important innovation in this instrument was its ability to clearly record the intervals during which current was applied to the line. It used Morse's dot-and-dash method until Royal Engineering House, Vermont, US, introduced an alphabet printing telegraph widely used in the 1860s.
46. Home Department, Public Proceedings, 23 April 1852, no. 13. Abstract. NAI.
47. M. Gorman, 'Sir William O'Shaughnessy, Lord Dalhousie, and the Establishment of the Telegraph System in India', *Technology and Culture*, vol. 12 (4), 1971, pp. 581–601.
48. *General Report of the Indian Telegraph Department for 1858–9*, Calcutta: Government Printing, 1859, pp. 541–2. NL.
49. *Report on the Operations of the Electric Telegraph in India, for the Last Quarter of 1855–6*, Calcutta: Government Printing, Appendix. NL
50. W. B. O'Shaughnessy, *Instructions Relative to Instruments and Offices for the Indian Telegraph Lines*, London, 1853. VMML.
51. Home Department, Public Proceedings, 21 June 1850, no. 28, from the Military Board to Major General John Hunter Littler, 4 June 1850. NAI.
52. *Report for the Last Quarter of 1855–6*, p. 2. NL
53. Home Department, Public Proceedings, 23 April 1852, no. 13. Abstract. NAI.
54. V/24/4282. *First Report on the Operations of the Electric Telegraph Department in India from 1February 1855 to 31 January 1856*, Calcutta: Thos. Jones, 1856; no. 350, from W. B. O'Shaughnessy, Superintendent, Electric Telegraphs, to C. Beadon, Secretary, Home Department, Government of India, 3 September 1855. India Office and Oriental Collection [Henceforth OIOC].
55. Ibid., pp. 36–7. OIOC.
56. Ibid., p. 7. OIOC.
57. Ibid., p. 4. OIOC.
58. V/24/4282. *Annual Report of the Telegraph Department 1859–1860*. Appendix. Abstract statement showing the total cash receipts, and pro-forma charges of each month, on account of Paid and Service messages transmitted by electric telegraph during the year 1856–7; also the total number of messages sent by 'Natives'. Compiled by Sheeb Chunder Nundee, In Charge, Office of the Officiating Superintendent, Electric Telegraph in India. OIOC.
59. Kalidas Maitra, *Electric Telegraph ba Taritbartabaha Prakaran*, Srirampur: J. H. Peters, 1855.
60. Kalidas Maitra, *Three Speeches by H. Pratt, Ramlochan Ghose and K. Moitre*, Calcutta, 1856.
61. Kalidas Maitra, *Muktavali Nataka*, Calcutta, 1857.
62. Kalidas Maitra, *The Steam Engine and the East India Railway Containing a History of India, with a Chronological Table of the Indian Princes from Judister Down to the Present Time etc.*, Srirampur, 1855.

63. Kalidas Maitra, *Manabdehotatwa or the Human Frame: An Easy and Familiar Introduction to the Principles of Anatomy and Physiology*, Srirampur, 1857.
64. Kalidas Maitra, *Paunarbhava Khandana: A Refutation of Isvarachandra Vidyasagar's Treatise in Favour of Widow Remarriage*, Srirampore, 1855.
65. Kalidas Maitra, *Vaspiyakal o Bharatiya Railway: The Steam Engine and the East India Railway Containing a History of India, with a Chronological Table of the Indian Princes, from Judister down to the Present Time with a Description of the Places and their Histories through which the Railway Passes, with a Coloured Map and Many Illustrations*, Srirampore: J. H. Peters, 1855.
66. Ibid., pp. 4–5.
67. Ibid., Introduction.
68. Ibid., p. 48.
69. Ibid., pp. 6–7.
70. Ibid., p. 3.
71. Kalidas Maitra, *Electric Telegraph ba Taritbartabaha Prakaran*.
72. Ibid., pp. 120–31.
73. Ibid., pp. 135–49.
74. Ibid., pp. 142, 146.
75. Ibid., pp. 141–5 *passim*.
76. Akshoy Kumar Dutta, *Directions for a Railway Traveller: Bashpiya ratharohidiger prati updesh; arthat jahara kaler gari arohon kariya gaman karen, tahader tatsankranta bighna nibaraner upay pradarshan*, Calcutta: Tattobodhini press, 1855.
77. Cf. Abdul Bismillah (ed.), *Bharatenduyugeen Vyanga*, Delhi: Sandarv Prakashan, 1989. I thank Dr Francesca Orsini for lending me this book and for other references.

2 The Telegraph and the Uprisings of 1857

1. Cf. E. Stokes, *The Peasant and the Raj: Studies in Agrarian Society and Peasant Rebellion in Colonial India*, Cambridge: Cambridge University Press, 1978.
2. P. V. Luke, 'How the Electric Telegraph Saves India', *Macmillan's Magazine*, vol. LXXVI, 1897, October 1897, pp. 401–6.
3. There were no opportunities for the registration of patents except for those inventions made on British soil; cf. Lahiri Choudhury, 'O'Shaughnessy and the Telegraph in India, c. 1836–1856'.
4. *Administration Report of the Telegraph Department for the years 1862–3, 1863–4, 1864–5 and 1865–6*, Calcutta, 1866. From Lieutenant Colonel D. G. Robinson, Director General Telegraphs, to E. C. Bayley, Secretary, Government of India, Home Department, 22 April 1866, p. 5. NL. Luke's article incorrectly attributes this statement to 1876: probably a misprint, see P. V. Luke, 'Early History of the Telegraph in India', *Journal of the Institute of Electrical Engineers*, vol. XX, 1891, pp. 102–22, CUL.
5. Luke, 'Early History', p. 105. CUL.
6. C. C. Adley, *The Story of the Telegraph in India*, London, 1866, p. 18.
7. Cf. *Administration Report of the Telegraph Department for the years 1862–3, 1863–4, 1864–5 and 1865–6*, Calcutta: Government Printing Press, 1866. From Lieutenant Colonel D. G. Robinson, Director General Telegraphs, to E. C. Bayley, Secretary, Government of India, Home Department, 22 April 1866. NL.

8. A. Sattin (ed.), *An Englishwoman in India: The Memoirs of Harriet Tytler 1828–1858*, Oxford and New York: Oxford University Press, 1986, p. 133.
 9. *General Report of the Indian Telegraph Department for 1858–9*, Calcutta: Government Printing, pp. 541–2. NL.
10. Home Department, Public Proceedings, 29 April 1853, no. 103. Minute by Lord Dalhousie, 29 April 1853. NAI.
11. Edney, *Mapping an Empire*, pp. 333–5.
12. What has been termed a 'honeycomb-pattern of differentiated routes connecting local and regional markets', R. Ahuja, *Pathways of Empire; Circulation, 'Public Works' and Social Space in Colonial Orissa, c. 1780–1914*, Hyderabad: Orient Blackswan, 2009, p.57; also Lahiri Choudhury, 'O'Shaughnessy and the Telegraph in India, c. 1836–1856'.
13. J. Deloche, *Transport and Communications in India Prior to Steam Locomotion*, (trans. by James Walker) 2 vols, Delhi: Oxford University Press, 1993–4.
14. Home Department, Public Proceedings, 23 April 1852, no. 14. Note by Lord Dalhousie, 14 April 1852. NAI.
15. Adley, *The Story of the Telegraph*, p. 8.
16. Cf. Lahiri Choudhury, 'Communication and Empire'.
17. V/24/4283, *Annual Report for 1860–1*. From Lieutenant Colonel C. Douglas, p. 27. OIOC.
18. Home Department, Public Proceedings, Electric Telegraph (A), nos 8–12, 6 March 1857, registered telegram no. 40, from G. T. Edmonstone, Secretary, Government of India, to Lieutenant L. Anderson, Government of Bombay, Calcutta, 25 November 1856. NAI.
19. Ibid., *Bombay Telegraph Courier*, 2 December 1856. NAI.
20. On the international response to 1857, see P. C. Joshi (ed.), *Rebellion: 1857–A Symposium*, Delhi, 1957; Yu Sheng-Wu and Chang Chen-Kun, 'China and India in the Mid-Nineteenth Century', pp. 332–53; also P. Shastiko, 'Russian Press on 1857', p. 332; Liliana Dalle Nogare, 'Echoes of 1857 in Italy', pp. 322–31; C. Fournian, 'Contemporary French Press', pp. 313–21. The paper on China compared 1857 and the Second Opium War 1856–60.
21. O'Shaughnessy, *First Report*, Appendix C. NAI.
22. Home Department, Public Proceedings, Electric Telegraph (A), nos 8–12, 6 March 1857. Blacknight's deposition, no. 312 of 1856, from J. Blacknight, Officiating First Class Inspector, Electric Telegraph, to R. L. Brunton, Deputy Superintendent, Pune, 4 December 1856. NAI.
23. Home Department, Public Proceedings, Electric Telegraph (A), nos 8–12, 6 March 1857. No. 30, from J. Blacknight, Officiating First Class Inspector, Electric Telegraph, to R. L. Brunton, Deputy Superintendent, Pune, 14 December 1856. NAI.
24. Home Department, Public Proceedings, Electric Telegraph (A), nos 8–12, 6 March 1857. No. 476, from C. Beadon, Secretary, Government of India, to Lieutenant P. Stewart, officiating Superintendent, Electric Telegraph in India, Calcutta, 6 March 1857. NAI.
25. Home Department, Public Proceedings, Fort William, no. 5, 16 October 1857. From Thomas Ogilvy, Magistrate of Dharwar, to W. Hart, Secretary to the Government of Bombay. No. 894, citing several instances of information incontinence, 4 June 1857. NAI.
26. Dosabhoy Framjee, *The British Raj Contrasted with its Predecessors and an Inquiry into the Disastrous Consequences of the Rebellion in the North-West*

Provinces upon the Hopes of the People of India, [Manager of the *Bombay Times*], London: Smith, Elder and Co., 1858. *Indian Tracts 1849–69*, Trinity College Library, Cambridge University [Henceforth TCL].

27. Anonymous, *Bengal Massacre!*, London: printed for private circulation, n.d.; letter published in the *Harkaru*, 5 September 1857. NL.
28. P. J. O. Taylor (gen. ed.), *A Companion to the 'Indian Mutiny' of 1857*, Delhi: Oxford University Press, 1996, p. 299.
29. *General Report of the Telegraph Department for 1857–8*. From O'Shaughnessy to C. Beadon, Secretary, Government of India. Calcutta: Government Printing Press, 1859, p. 4. NL.
30. V/24/4282. *Annual Report for 1859–60*, p. 3. OIOC.
31. *General Report for 1857–8*, p. 3. NL.
32. Lahiri Choudhury, 'O'Shaughnessy and the Telegraph in India', pp. 331–59.
33. M. R. Gubbins, *The Mutinies in Oudh*, London, 1858 (rep. Patna: Janaki Prakashan, 1978), p. 360.
34. *General Report for 1857–8*, p. 4. NL.
35. Home Department, Public Proceedings, Electric Telegraph, 14 August 1857, nos 5–6. From the Acting Superintendent of the Telegraph, to Secretary, Government of India, 15 July 1857. NAI.
36. *General Report for 1857–8*, p. 5. NL.
37. Iltudus Thomas Pritchard, *The Mutinies in Rajpootana*, London, 1860 (repr.1976), p. 192.
38. Gubbins, *The Mutinies in Oudh*, pp. 380–1.
39. *The Indian Post and Telegraph Magazine*, vol. I, January 1920, no. 1, p. 50. NL.
40. *The Indian Post and Telegraph Magazine*, vol. I, March 1920, no. 3, p. 20. NL.
41. Taylor, *A Companion to the 'Indian Mutiny'*, p. 327.
42. S. A. A. Rizvi and M. L. Bhargava (eds), *Freedom Struggle in Uttar Pradesh*, vol. II, Uttar Pradesh: Publications Bureau, 1958, pp. 10–11.
43. Bayly, *Information and Empire*, pp. 319–22.
44. J. C. Marshman, *Memoirs of Major General Sir Henry Havelock*, London, 1860, p. 175.
45. Bayly, *Information and Empire*, pp. 319–22.
46. Home Public Proceedings, Electric Telegraph, nos 6–7, 4 December 1857. From Lieutenant Colonel R. Strachey, Secretary, Central Provinces, to C. Beadon, Secretary, Home Department, Government of India, no. 866, 9 October 1857. NAI.
47. Home Public Proceedings, Post Office, no. 14, 30 October 1857. From E. Maltby, Acting Chief Secretary to the Govt. of Bombay, to C. Beadon, Secretary, Home Department, Government of India, no. 1334, 21 September 1857. NAI.
48. Home Public Proceedings, Post office, no. 7, 30 October 1857. Endorsement no. 1119, from the Junior Secretary, Government of Bengal, forwarding a letter from the Judge of Jessore, 8 October 1857. NAI.
49. Ibid. NAI.
50. Home Public Proceedings, Fort William, no. 4, 6 September 1857. From Captain C. Holroyd, Principal Assistant Commissioner, to Colonel Jenkins, Agent to the Governor General, North East Province. NAI.
51. Home Public Proceedings, Fort William, no. 21, 30 October 1857. From C. Beadon, Secretary, Government of India, to Sir W. B. O'Shaughnessy, no. 1747, 29 October 1857. NAI.

52. Home Public Proceedings, nos 12 and 13, 4 September 1857. Despatch from the Court of Directors, 8 July 1857. NAI.
53. Copies were also sent to the governments of Bombay and Madras and the Chief Commissioner of the Punjab; Home Public Proceedings, no. 15, 4 September 1857. NAI.
54. Home Public Proceedings, Fort William, no. 10, 30 October 1857. From C. Beadon to E. Maltby, Secretary, Government of Bombay, no. 1171. NAI.
55. Literally, language of the night or an evening language.
56. *Delhi Gazette*, 24 August 1857, D. E. Augier papers, Major C. E. Luard (comp.), 'Contemporary Newspaper Accounts of Events During the Mutiny in Central India, 1857–9', typescript mss, p. 16. Centre for South Asian Studies [Henceforth CSAS].
57. Home Public Proceedings, Fort William, no. 6, 23 October 1857. From R. Knight, ed., *Bombay Times*, to H. L. Anderson, Secretary, Government of Bombay, 24 September 1857. NAI.
58. Home Public Proceedings, Fort William, no. 6, 23 October 1857. No. 3235, from H. L. Anderson in reply, 26 September 1857. NAI.
59. Home Public Proceedings, Fort William, no. 68, 23 October 1857. Message from Agra, 2 October 1857. NAI.
60. George O. Trevelyan, *Cawnpore*, London: Macmillan and Co., 1894 (4th edn), p. 53.
61. Council of India Minutes and Memoranda 1858–1947: C/137/ 20–79ff. India Office, 30 April 1874. Lieutenant Colonel Owen Tudor Burne. Historical Summary of the Central Asian Question. Appendix VII, 76ff. (Translated by J. Fred. Hodgson, Lieutenant and Interpreter, H. M. Bengal Army). OIOC.
62. Ibid. OIOC.
63. Trevelyan, *Cawnpore*, p. 55.
64. V. D. Savarkar, *The Indian War of Independence*, Delhi, 1909, p. 82.
65. Taylor, *A Companion*, p. 304.
66. Home Public Proceedings, Fort William, no. 66, 23 October 1857. From Durgaprasad Roy Chowdhury, Bhowanipur, to the Secretary, Home Department, 14 October 1857. NAI.
67. Cf. M. H. Fisher, 'The Office of the Akhbar Nawis: The Transition from Mughal to British Forms', *Modern Asian Studies*, 27 (1) 1993, pp. 45–82.
68. Rizvi and Bhargava, *Freedom Struggle in Uttar Pradesh*, pp. 7–8.
69. Trevelyan, *Cawnpore*, p. 98.
70. Augier papers. Appendix A: The History of the Murder of Major Neill at Augur in 1887, p. 6. CSAS.
71. Luke, 'Electric Telegraph Saves India', p. 406.

3 The Discipline of Technology

1. On the process of standardisation cf. Simon Schaffer, 'A Manufactory of OHMS, Victorian Metrology and Its Instrumentation', in R. Bud and S. Cozzens (eds), *Invisible Connections*, Bellingham: SPIE Optical Engineering Press, 1992, pp. 25–54.
2. Cf. R. A. Buchanan, 'Institutional Proliferation in the British Engineering Profession, 1847–1914, *The Economic History Review*, New Series: vol. 38 (I), February 1989, pp. 42–60.

3. *General Report of the Telegraph Department for 1855–6 to 1859–60*, Calcutta: Government Printing, 1860. NL; cf. *General Report of the Telegraph Department for 1857–8*, from Dr W. B. O'Shaughnessy, Superintendent, Telegraphs, to Cecil Beadon, Secretary, Government of India, forwarded with appendices by Lieutenant P. Stewart, Deputy Superintendent, Telegraphs, 8 April 1858, p. 9. NAI.
4. M. H. Fisher, 'The East India Company's Suppression of the Native Dak', *Indian Economic and Social History Review*, xxxi (3), 1994, pp. 319–26.
5. Please see Chapter 2.
6. *General Report for 1859–1860*. From W. B. O'Shaughnessy, Superintendent, Telegraphs, to W. Grey, Secretary, Government of India, no. 303, 17 May 1860. NAI.
7. Iwan Rhys Morus, *Frankenstein's Children: Electricity, Exhibition, and Experiment in Early Nineteenth Century London*, Princeton; Chichester: Princeton University Press, 1998.
8. *Administration Report for 1862–3, 1863–4, 1864–5 and 1865–6*, Calcutta, 1866. From Lieutenant Colonel D. G. Robinson, Director General Telegraphs, to E. C. Bayley, Secretary, Government of India, Home Department, 22 April 1866, p. 5. NL.
9. V/24/4282. *First Report*, p. 9. OIOC.
10. Ibid., no. 417, from O'Shaughnessy to C. Allen, Officiating Secretary, Home Department, Government of India, 4 May 1856. OIOC.
11. Ibid., p. 8. OIOC.
12. V/24/4282. *Annual Report 1859–1860*, p. 12. OIOC.
13. V/2/4284. *Annual Report for 1861–2*, Appendix D. From A. M. Monteath, Under Secretary, Home Department, Government of India, to Lieutenant Colonel C. Douglas, no. 2154, 30 November 1861, transmitting memorial by Sir W. B. O'Shaughnessy on the subject of the re-organisation of the Indian Telegraph Department, Liverpool, 15 August 1861. OIOC.
14. *Report of the Joint Committee appointed by the Lords of the Committee of Privy Council for Trade and the Atlantic Telegraph Company to Inquire into the Construction of Submarine Telegraph Cables; Together with Minutes of Evidence and Appendix*, Parliamentary Papers, vol. LXII, 1860. OIOC.
15. *Report on the Existing State of Telegraph Communication in India and the Projects for its Extension*, Calcutta, 1864, p. 10; from Lieutenant Colonel C. Douglas, Director General of Telegraphs in India, to Secretary, Government of India, no.3890, 8 April 1864, Calcutta: Government Printing, 1864. NL.
16. Ibid., p. 11. NL
17. *Annual Report of the Telegraph Department 1859–1860*, p. 9. NL.
18. Ibid., p. 10. NL.
19. *Annual Report for 1866–1871*, p. 21. NL.
20. Ibid., Appendix L, p. 50. NL.
21. Ibid., p. 22. NL.
22. *Administrative Report of the Indian Telegraph Department for 1886–7*, Simla, 1888, p. 10. NL.
23. *Annual Report 1866–1871*, p. 20. NL.
24. Ibid., p. 52. NL.
25. Anonymous, *The Late Government Bank of Bombay: Its History*, London: Private Printing, 1868, p. 15; *Indian Tracts 1849–69*. TCL.

26. Storey, *Reuters' Century*, pp. 62–3.
27. Ibid., pp. 43–4. TCL.
28. Ibid., p. 45. TCL.
29. Ibid., p. 53. TCL.
30. Mss. Eur. A130. Letter, 22 April 1872, from J. S. Mill to Edwin Arnold, Leader-writer of the *Telegraph*, discussing the role of the Bombay Government in the failure of the Bank of Bombay in 1866. OIOC. Private Papers. For a comprehensive account, see *Report of the Commission on the Failure of the Bank of Bombay, Parliamentary Papers*, 1868–9 (4162) xv. OIOC.
31. M. Vicziany, 'Bombay Merchants and Structural Changes in the Export Community 1850 to 1880', in K. N. Chaudhuri and C. J. Dewey (eds), *Economy and Society: Essays in Indian Economic and Social History*, Delhi: Oxford University Press, 1979, pp.163–96.
32. Storey, *Reuters' Century*, p. 63.
33. Ibid., p. 64.
34. V/24/4282. *First Report for 1855–56*, no.1, from O'Shaughnessy, Superintendent, Telegraphs, to C. Beadon, Secretary, Home Department, Government of India, 9 February 1856, Enclosing the report for 1856. OIOC.
35. V/2/4283, *Annual Report for 1860–1*, p. 16. OIOC.
36. Ibid., p. 21. OIOC.
37. V/24/4282, *First Report*, p. 59. OIOC.
38. V/2/4283, *Annual Report for 1860–1*, Appendix B. OIOC.
39. V/24/4282, *First Report*, Appendix. No. 292, from S. Nundee, Assistant in charge of Benares Division, to Sir W. B. O'Shaughnessy, 22 September 1855. OIOC.
40. V/2/4284, *Annual Report for 1861–2*, Appendix D. OIOC.
41. V/24/4282, *.Annual Report for1859–1860*, p. 12. OIOC.
42. Ibid. OIOC.
43. V/24/4282, *First Report*, p. 61. OIOC.
44. Ibid., p. 78. OIOC.
45. V/24/4282, *First Report*; *Annual Report for1859–60*. OIOC.
46. V/24/4284, *Annual Report for 1861–2*, p. 35. OIOC.
47. Ibid.
48. Ibid., Appendix. General Branch Circular, no. 96 (3/4), 25 February 1861. OIOC.
49. Ibid., p. 32. OIOC.
50. Home Public Proceedings, Electric Telegraph (A), nos 8–12, 6 March 1857. Blacknight's deposition, no. 312 of 1856, from J. Blacknight, Officiating First Class Inspector, Electric Telegraph, to R. L. Brunton, Deputy Superintendent, Pune, 4 December 1856. NAI.
51. V/24/4284. *Annual Report for 1861–2*, Appendix. General Branch Circular, no. 98 (1/2), 1 March 1864. OIOC.
52. *Council of India: Minutes and Memoranda 1858–1947*. C/138 131–3ff. G. Chesney, Coopers Hill, 3 November 1875. Memorandum regarding some of the conditions of service of Civil Engineers in the Indian Public works Department. OIOC.
53. C/138 131–3ff. OIOC.
54. *Council of India: Minutes and Memoranda 1858–1947*. C/142 395–9ff. Sir George Tomkyns Chesney, Memorandum on Cooper's Hill and the supply of engineers for India, 30 May 1879. OIOC.

55. *Council of India: Minutes and Memoranda 1858–1947*, C/142 395–9ff., ff. 397. OIOC.
56. V/2/4283, *Annual Report for 1860–1*, p. 28. OIOC.
57. *Annual Report of the Telegraph Department 1861–2*, p.16. NL.
58. V/2/4283, *Annual Report for 1860–1*, Appendix B. From W. Grey, Secretary, Home Department, Government of India, to the Secretary, Bengal Chamber of Commerce, 12 June 1861. Similar letters were addressed to the other Chambers including Bombay and Madras Chambers of Commerce. OIOC.
59. Ibid. From H. W. J. Wood, Secretary, Bengal Chamber of Commerce, to Secretary, Home Department, Government of India, with enclosures, 14 August 1861. OIOC.
60. V/24/4289, *Administrative Reports of the Indo-European Telegraph Department, 1872–3 to 1894–5. Report for 1872–3*, p. 3. OIOC.
61. V/24/4289, *Report for 1872–3*, p. 7. OIOC.
62. V/2/4283, *Annual Report for 1860–1*, p. 14. OIOC.
63. *Annual Report of the Telegraph Department 1859–1860*, Calcutta, 1860, p. 10. NL.
64. *Annual Report of the Telegraph Department 1866–7 to 1870–1*, Calcutta, 1871. From Colonel D. G. Robinson, Director General, Telegraphs, sent through the Secretary, Government of Bombay, 24 July 1871, Calcutta: Government Printing, p. 47. NL.
65. Geoffrey Bennington, 'Postal Politics and the Institution of the Nation', in Homi K. Bhabha (ed.), *Nation and Narration*, London: Routledge, 1990, pp. 121–37.
66. *The Telegram and Telegrapheme Controversy, as Carried on in a Friendly Correspondence Between A, C. and H., both M A's of Trinity College, Cambridge*, London: Rivingtons, 1858 (Of course, the exchange was anything but friendly); *Tracts: Cambridge 1848 to 1885*. TCL. I thank Dr A. R. Cox for his help with accessing the *Tracts*.
67. A. Brasher, *Telegraph to India: Suggestions to Senders of Messages*, London: Edward Stanford, 1870, *Scientific Tracts 1–13*. pp. 7–9. OIOC.
68. V/2/4284, *Annual Report of the Telegraph Department 1861–2*, p. 15. OIOC.
69. Brasher, *Telegraph to India*, p. 9. OIOC.
70. V/2/4284, *Annual Report 1861–2*. Appendix J. General Branch Circular no. 110. From Lieutenant C. Douglas, to all offices, 1 July 1861. OIOC.
71. Brasher, *Telegraph to India*, p. 13. OIOC.
72. *Annual Report for1861–2*, Appendix D. From A. M. Monteath, Under Secretary, Home Department, Government of India, to Lieutenant Colonel C. Douglas, no.2154, 30 November 1861, transmitting memorial by Sir W. B. O'Shaughnessy on the subject of the re-organisation of the Indian Telegraph Department, Liverpool, 15 August 1861. NL.
73. D. T. Ansted, *The Bottom of the Atlantic and the First Laying of the Telegraph Cable: Being the Substance of a Lecture Delivered at the Working Men's Association of Guernsey*, Guernsey: F. Le lievre, 1860, p. 23; also W. H. Barlow, 'On the Spontaneous Electrical Currents Observed in the Wires of the Electric Telegraph', *Philosophical Transactions of the Royal Society of London*, vol. 139, (1849), pp. 61–72.
74. Lester G. Lindley, *The Impact of the Telegraph on Contract Law*, New York and London: Garland Publishing Inc., 1990, p. 31.

75. Jeffrey Kieve, *The Electric Telegraph: A Social and Economic History*, Newton Abbot: David & Charles, 1973, p. 249.
76. *Souvenir of the Inaugural Fete at Mr Pender's 18 Arlington Street to Celebrate the Opening of the Direct Submarine Telegraphic Communication to India on 23rd of June 1870*, London: Private Printing, 1870, p. 50. OIOC.
77. V/24/4282, Annual Report for 1859–1860, p. 12. OIOC.
78. Government of the Union of South Africa, *Speedways of Thought: The Romance of the Development of the Telephone, the Telegraph, of Wireless Telephony and the Air Mail*, Pretoria: Government Printer, 1937, p. 35. OIOC.
79. Irvin Stewart, 'The International Telegraph Conference of Brussels and the Problem of Code Language', *American Journal of International Law*, Volume 23, Issue 2 (April 1929), pp. 292–306.
80. Ibid.

4 Making the Twain Meet: The New Imperialism of Telegraphy

1. W. Rodney, *How Europe Underdeveloped Africa*, Washington: Howard University Press, 1982; A. Gunder Frank; *Capitalism and Underdevelopment in Latin America*, New York: Monthly Review Press, 1967.
2. Cf. E. Stokes, 'Late Nineteenth Century Colonial Expansion and the Attack on the Theory of Economic Imperialism: A Case of Mistaken Identity?' *The Historical Journal*, vol. XII, 2 (1969), pp. 285–301, p. 286.
3. Ronald Robinson and John Gallagher, 'The Imperialism of Free Trade, 1815–1914', *The Economic History Review*, Second Series, vol. VI, no. I (1953) pp.1–15.
4. For qualifications, see P. J. Cain, A. G. Hopkins, 'Gentlemanly Capitalism and British Expansion Overseas II: New Imperialism, 1850–1945', *The Economic History Review*, New Series, vol. 40, I (February 1987), pp. 1–26, p.2, 4, 6, *passim*.
5. A. Seal, *The Emergence of Indian Nationalism: Competition and Collaboration in the Later Nineteenth Century*, Cambridge: Cambridge University Press, 1968.
6. Cf. Richard Shannon, *The Crisis of Imperialism 1865–1915*, London: Granada, 1976, p. 219.
7. D. K. Fieldhouse, '"Imperialism": An Historiographical Revision', *The Economic History Review*, Second Series, vol. XIV, no. 2, 1961, pp. 187–209, pp. 201–2.
8. D. K. Fieldhouse, *Economics and Empire 1830–1914*, London: Widenfeld and Nicholson, 1973.
9. Cf. D. Headrick, 'The Tools of Imperialism: Technology and the Expansion of European Colonial Empires in the Nineteenth Century', *The Journal of Modern History*, vol. 51, 2 (June 1979), pp. 231–63.
10. C. Cipolla, *Guns and Sails in the Early Phase of European Expansion 1400–1700*, London: Collins, 1965.
11. P. J. Marshall, 'Western Arms in Maritime Asia in the Early Phases of Expansion', *Modern Asian Studies*, vol. XIV, 1980, pp. 13–28.
12. Cf. H. L. Wesseling, *Imperialism and Colonialism: Essays on the History of European Expansion*, Contributions in Comparative Colonial Studies No. 32, London: Greenwood, 1997, pp. 27–37.

13. Headrick, 'The Tools of Imperialism', pp. 233–4.
14. Ibid., pp. 259–61.
15. Ibid., p. 260. Lord Kelvin used a similar sounding phrase: 'armed science'. I discuss this idea in more detail later on in this book.
16. Ibid., p. 259; W. Churchill, *The River War: An Account of the Reconquest of the Soudan*, (abridged edition) London: Eyre and Spottiswoode, 1933, p. 274.
17. Robinson, 'Non-European Foundations of European Imperialism' in E. R. J. Owen and R. Sutcliffe (eds), *Studies in the Theories of Imperialism*, London: Longman, 1972, p. 132.
18. Clarence B. Davis, Kenneth E. Wilburn, Jr (eds), with Ronald E. Robinson, *Railway Imperialism* (*Contributions in Comparative Colonial Studies*), no. 26, London: Greenwood, 1991, p. 2.
19. Hugh Barty-King, *Girdle Round the Earth: The Story of Cable and Wireless and its Predecessors to Mark the Group's Jubilee 1929–1979*, London: Heinemann, 1979, p. 37.
20. Roderic H. Davison, 'Effect of the Electric Telegraph on the Conduct of Ottoman Foreign Relations', in Caesar E. Farah (ed.), *Decision Making and Change in the Ottoman Empire*, Kirksville: Thomas Jefferson University Press, 1993, pp. 53–66.
21. Charles Bright, *The Life Story of Sir Charles Tilston Bright*, London: A. Constable, 1908, p. 219.
22. George Balfour, *Trade and Salt in India Free; Reprints (by permission) of Articles and Letters from the Times*, London, 1875, pp. 89–90, *Indian Tracts 1849–69*. TCL.
23. Ibid., pp. 35–6.
24. Cf. P. Hopkirk, *The Great Game: On Secret Service in High Asia*, Oxford: Oxford University Press, 1990.
25. Daniel R. Headrick, *The Tentacles of Progress: Technology Transfer in the Age of Imperialism, 1850–1940*, New York: Oxford University Press, 1988, p. 14.
26. Colonel Sir F. J. Goldsmid, *Telegraph and Travel: A Narrative of the Formation and Development of Telegraphic Communication between England and India etc*, London: R. Clay, Sons, and Taylor, 1874, p. 23.
27. Electric Telegraph, vol. I, 1856–9, Letters to and from the Court of Directors, Government of Bombay, no. 6, May 1859. Copy of a letter addressed to Brigadier Coghlan, 3 May 1859. Maharashtra State Archives [Henceforth MSA].
28. Ibid., no. 4, February 1859. From the Red Sea Company to the Court, 11 February 1859. MSA.
29. Ibid., no. 5, July 1859, From the Red Sea Company to the Court, 30 June 1859. MSA.
30. Ibid., no. 17, August 1859, Copy of correspondence with the Red Sea and India Telegraph Company. From the Red Sea Company, 19 July 1859. MSA.
31. Ibid., no. 13, August 1859, From the Red Sea Company, 19 August 1859. MSA.
32. Ibid., no. 9, August 1859, Governor-General-in-Council, Bombay to the India Office, 4 August 1859. MSA.
33. Ibid., no. 11, September 1859, From the Red Sea Company to the Court, 30 August 1859. MSA.
34. Ibid., no. 13, July 1859, From the Red Sea Company to the Court, 23 June 1859. MSA.
35. Ibid., no. 25, November 1859, From Brigadier Coghlan, Consul General, Egypt, to Lord John Russell, 20 October 1859. MSA.

36. Ibid., no. 16, November 1859, From the Red Sea Company to the Court, 17 November 1859. MSA.
37. Anonymous, *The Story of my Life by the Submarine Telegraph*, London: C. West, 1859, p. 95. OIOC.
38. Ibid., p. 3. OIOC.
39. Ibid., p. 89. OIOC.
40. Ibid., pp. 83–91. OIOC.
41. *Annual Report of the Telegraph Department 1866–7 to 1870–1*, p. 3. NL.
42. *Souvenir*, p. 15. WPL.
43. Coates, Finn et al., *A Retrospective Technology Assessment*, p. vii.
44. Goldsmid, *Telegraph and Travel*, pp. 81–2.
45. Ibid., p. 23.
46. Mss Eur. C168/B. Colonel Sir F. J. Goldsmid's Diary for 1864, 21 January 1864. OIOC.
47. Ibid., 9 January 1864. OIOC.
48. Goldsmid, *Telegraph and Travel*, p. 105.
49. Mss Eur. C168/B. Goldsmid, Diary, 9 January 1864. OIOC.
50. Public Works Department, Telegraph Establishment (A) Proceedings, March 1871, nos 3–10. No. 7, from Colonel D. G. Robinson, Director General, Indian Telegraph Department, to Secretary, Public Works Department, Government of India, no. 451, 14 November 1870. NAI.
51. Goldsmid, *Telegraph and Travel*, p. 137.
52. Government of India, Mekran Telegraph Department, 1863–4. Letters to the Home Department, Government of India, no. 14, 1864. From Colonel A. B. Kemball, Political Agent in Turkish Arabia to Secretary, Government of Bombay, Political Department, Baghdad, 15 February 1864. MSA.
53. Mss Eur. C168/B. Goldsmid, Diary, 30 January 1864. OIOC.
54. Goldsmid, *Telegraph and Travel*, p. 134.
55. Edney, *Mapping an Empire*, pp. 333–5.
56. Mss Eur. C168/B. Goldsmid, Diary, 1864. OIOC.
57. Letter from Colonel Sankey to Consul Skene at Aleppo, 29 June 1857, cited in Goldsmid, *Telegraph and Travel*, p. 85.
58. Foreign Department, Political (A), Governor General's proceedings, August 1863, nos 32 and 34. Proceedings in council. NAI.
59. Foreign Department, Political (A), May 1863, nos 52 and 56. Proceedings in council. NAI.
60. Foreign Department, Political (A), July 1863, nos 44 and 45. Proceedings in council. NAI.
61. Foreign Department, Political (A), September 1864, no. 99. From H. L. Anderson, Chief Secretary, Government of Bombay, to Secretary, Foreign Department, Government of India, 23 August 1864. NAI.
62. Goldsmid, *Telegraph and Travel*, pp. 86–7.
63. Mss Eur. C168/B. Goldsmid, Diary, 13 February 1864. OIOC; cf., Goldsmid, *Telegraph and Travel*, p. 141.
64. Goldsmid, *Telegraph and Travel*, pp. 142–3.
65. Mss Eur. C168/B. Goldsmid, Diary, Goldsmid, Diary, 25 January 1864. OIOC.
66. Ibid., 30 May 1864. OIOC; cf., Goldsmid, *Telegraph and Travel*, p. 85.
67. Ibid., 13 February 1864. OIOC.
68. Ibid., 14 February 1864. OIOC.
69. Ibid., 3 and 5 March 1864. OIOC.

70. Ibid., 7 March 1864. OIOC.
71. Ibid., 24 and 25 February 1864. OIOC.
72. Cf. Section on Arms Trade in the Persian Gulf.
73. Goldsmid, *Telegraph and Travel*, p. 142.
74. General Department, Indo-European Telegraph Department, no. 27, 1866, from Lieutenant Colonel F. J. Goldsmid, Superintendent, Indo-European Telegraph Department, to Secretary, Government of Bombay, 27 March 1866. MSA.
75. Goldsmid, Diary, 3 April 1864. OIOC.
76. Goldsmid, *Telegraph and Travel*, pp. 132–3.
77. Cf. Davison, 'Effect of the Electric Telegraph'.
78. Edward Granville Browne, *A Year amongst the Persians: Impressions as to the Life, Character, and Thought of the People of Persia, 1887–88*, London: Adam and Charles Black, 1926, p. 99.
79. Basil Stewart (ed.), Colonel Charles E. Stewart, *Through Persia in Disguise: With Reminiscences of the Indian Mutiny*, London: Routledge, 1911, p. 339.
80. Browne, *A Year amongst the Persians*, p. 467.
81. Cf. General Department, Indo-European Telegraph Department, no. 9, 1864, from DG, Indo-European Telegraph Department, to Secretary of State for India: statements of accompanying subordinate European and Indian Engineering Staff proposed to be entertained for the Persian Gulf Submarine section, 11 July 1866. MSA.
82. Goldsmid, Diary, 29 May 1864. OIOC.
83. Mss Eur. C732. E. M. Norris, 'A Career in Persia 1894–1930', Hugh Norris (ed.), typescript, p. 30, Norris Private Papers. OIOC.
84. *Souvenir of the Banquet and Evening Fete in Celebration of the 25th Anniversary of the Establishment of Submarine Telegraphy with the Far East*, London, 1894, p. 140. Whipple Library, History and Philosophy of Science Department, University of Cambridge [henceforth WPL].
85. Stewart, *Through Persia in Disguise*, p. 266.
86. Ibid., p. 281.
87. Ibid., pp. 286–7.
88. Ibid., p. 287.
89. Ibid., p. 282.
90. For example, Captain J. N. Price Wood, *Travel and Sport in Turkestan*, London: Chapman & Hall, 1910.
91. Ameen Rihani, *Around the Coasts of Arabia*, London: Constable Co., 1930.
92. F. B. Bradley-Birt, *Through Persia from the Gulf to the Caspian*, London: Smith Elder & Co., 1909, p. 59.
93. Goldsmid, *Telegraph and Travel*, pp. 56–9.
94. Browne, *A Year amongst the Persians*, pp. 469–70.
95. Stewart, *Through Persia in Disguise*, Appendix II, 'On the Use of Petroleum as Fuel in Steamships and Locomotives, based on its Employment in that way in the Caspian Sea and in the Trans-Caspian Region', *Proceedings of the Royal United Service Institution for June 1886*, pp. 410–19.
96. Cf. L. P. Elwell-Sutton, *Persian Oil: A Study in Power Politics*, London: Lawrence Wishart, 1955.
97. Stewart, *Through Persia in Disguise*, Appendix II, pp. 211–13.
98. Mss Eur. B150, letter from Campbell to 'Amroo' (Mrs Campbell?), 19 February 1898, Robert Charles Arthur Campbell Private Papers. OIOC.

99. Ibid., Letter from Campbell to 'Amroo', 27 January 1900. OIOC.
100. *Supplement to the Gazette of India*, no. 862, 28 August 1885. Enclosure to Public Works Department Resolution no. 21T of 1886, from J. U. Bateman-Champain, Director General, Indo-European Telegraph Department, to Secretary, Government of India, Public Works Department. Administrative Report of the Indo-European Telegraph Department for 1884–5. OIOC.
101. Mss Eur. F211, Letter from Sir W. M. N. Young to his Aunt Jessie, 6 May 1888, Meston and Young Private Papers. OIOC.
102. Government of India, Public Works Department, Civil Works, Telegraphs (A), May 1887, nos 24–9. No. 27, from A. J. Leppoc-Cappel, Director General, Indian Telegraphs, to Secretary, Government of India, Public Works Department, no. 405–T, 2 April 1887, Report on operations in Upper Burma. NAI.
103. Cf. Norris, 'A Career in Persia 1894–1930', p. 6. OIOC.
104. Hugh Norris, 'A Little Family History' in Norris, 'A Career in Persia 1894–1930'. OIOC.
105. Barty-King, *Girdle Round the Earth*, pp. 86–7.
106. Goldsmid, Diary, 23 March 1864. OIOC.
107. Norris, 'A Career in Persia 1894–1930', p. 65. OIOC.
108. D. C. M. Platt, 'The Imperialism of Free Trade: Some Reservations', *Economic History Review*, new series, vol. 21, no. 2 (August 1968), pp. 296–306, p. 297.
109. Goldsmid, Diary, 20 February 1864. OIOC.
110. Stewart, *Through Persia in Disguise*, pp. 246–7.
111. Dharani Kanta Lahiri Choudhury, *Bharat-Bhraman*, Calcutta: Kuntolin Press, 1908, pp. 84–8. CUL.
112. Stewart, *Through Persia in Disguise*, p. 94.
113. Ibid., p. 96.
114. Ibid., pp. 264–5.
115. Cf. Davison, 'Effect of the Electric Telegraph', p. 56.
116. Stewart, *Through Persia in Disguise*, pp. 95–7.

5 The Magical Mystery Tour: Cable Telegraphy

1. *Souvenir of the Banquet and Evening Fete in Celebration of the 25th Anniversary of the Establishment of Submarine Telegraphy with the Far East held at the Imperial Institute, London, on Friday, 25th July 1894*, London: Private Printing, 1894. WPL.
2. J. Munro, *The Wire and the Wave: A Tale of the Submarine Telegraph*, The Boys Own Bookshelf Series, London: The Religious Tract Society, n.d., p. 33.
3. *Souvenir*, p. 8. WPL. It must be noted that Lord Kelvin was referring to the high degree of scientific achievement in the armed services; this period saw the army engineers replace the gentlemen of science in the colonies, cf. section on standardisation.
4. Ibid., p. 89. WPL.
5. Bruce J. Hunt, 'Doing Science in a Global Empire', p. 320.
6. Bright, *Sir Charles Tilston Bright*, p. 223.
7. Goldsmid, *Telegraph and Travel*, p. 1.
8. Ibid., pp. 81–2.

9. Headrick, *The Tools of Empire: Technology and European Imperialism in the Nineteenth Century*, New York: Oxford University Press, 1981.
10. George Basalla, 'The Spread of Western Science', *Science*, 156, pp. 611–21.
11. See Crosbie Smith and M. Norton Wise, *Energy and Empire: A Biographical Study of Lord Kelvin*, Cambridge: Cambridge University Press, 1989.
12. http://www.bbc.co.uk/dna/h2g2/A356889. Last visited 11/01/2006.
13. *Report on the Construction of Submarine Telegraph Cables*, Parliamentary Papers, vol. LXII, 1860. OIOC.
14. Standage, *The Victorian Internet*, p. 84.
15. Cf. Hunt, 'Doing Science in a Global Empire', pp. 312–33.
16. *Annual Report of the Telegraph Department 1858–9*, Calcutta, 1859. From W. B. O'Shaughnessy, Superintendent, Electric Telegraphs, to the Secretary, Government of India, 31 October 1858, p. 25. NL.
17. Cf. J. T. Walker, 'India's contribution to Geodesy', *Philosophical Transactions of the Royal Society of London*, vol. 186, 1895, pp. 745–816.
18. Institute of Civil Engineers, Minutes of proceedings, with abstracts of discussions, vol. 25, Session 1865–6, no. 1135, 14 November 1865, Sir Charles Tilston Bright, on the telegraph to India, and its extension to Australia and China, pp. 1–65. Institute of Civil Engineers [Henceforth ICE].
19. Secretariat Record Office, Indo-European Telegraph Department, letters to the Secretary of State for India, nos 19–20, from Government of Bombay, 3 June 1865. MSA.
20. Tracts, TFV50, folio 50, no. 6, report from F. C. Webb on the operations of picking up and repairing the submarine cable between Bushire and Fao, extending over a period from 29 March to 3 April 1864, to Lieutenant Colonel. P. Stewart, Director General, Indian Telegraph Department, 1 June 1864. ICE.
21. *Report of the Joint Committee appointed by the Lords of the Committee of the Privy Council for Trade and the Atlantic Telegraph Company to inquire into the Construction of Submarine Telegraph Cables*, HMSO 1861. OIOC.
22. *Europe and America: Reports of Proceedings at an Inauguration Banquet given the 15th April 1864, in Commemoration of the Renewal by the Atlantic Telegraph Company (after six years) of their Efforts to Unite Ireland and Newfoundland by means of the a Submarine Cable; and at an Anniversary Banquet given the 10th March 1868, in Commemoration of the Agreement for the Establishment of a Telegraph Across the Atlantic, on the 10th of March, 1854, by Mr. Cyrus W. Field*, printed for private circulation, n.d., p. 21. OIOC.
23. *Society of Telegraph Engineers*, proceedings volume, pp. 42–6; 9 February 1876, 43rd ordinary general meeting, C. V. Walker, President and Chair. GPO.
24. Ibid., pp. 81–3; 23 February 1876, 44th ordinary general meeting, C. V. Walker, President and Chair. GPO.
25. Barty-King, *Girdle Round the Earth*, p. 40.
26. Ibid., p. 79.
27. Daniel R. Headrick, *The Tools of Empire*, p. 29.
28. Baark, *Lightning Wires*; Ahvenainen, *The Far Eastern Telegraphs*; Ahvenainen, *The History of the Caribbean Telegraphs before the First World War*, Helsinki: Annales Academiae Scientiarum Fennicae. Ser. Humaniora, Finnish Academy of Science and Letters, vol. 283, 1996.

29. Daniel R. Headrick, *The Tentacles of Progress: Technology Transfer in the Age of Imperialism, 1850–1940*, New York: Oxford University Press, 1988, pp. 97–144.
30. For details see, Vary T. Coates, Bernard Finn et al., *A Retrospective Technology Assessment: Submarine Telegraphy; The Transatlantic Cable of 1866*, San Francisco: San Francisco Press Inc., for the George Washington University Program of Policy Studies in Science and Technology, 1979.
31. Barty-King, *Girdle Round the Earth*, p. 15.
32. Headrick, *The Tentacles of Progress*, p. 102.
33. Kieve, *The Electric Telegraph*, p. 117.
34. J. Wagstaff Blundell, *The Manual of Submarine Telegraph Companies*, London: Published by the Author, 1872 edition, p. 27. OIOC.
35. Ibid. OIOC.
36. Headrick, *The Tentacles of Progress*, p. 103.
37. Barty-King, *Girdle Round the Earth*, p. 38.
38. Ibid., p. 39.
39. Ibid., p. 46; cf. Platt, 'Imperialism of Free Trade: … Reservations', p. 299.
40. Ibid., p. 23.
41. Ibid., p. 52.
42. Cf. Gore Vidal, *Daily Mirror*, 10 July 2002.
43. Menahem Blondheim, *News Over the Wires: The Telegraph and the Flow of Public Information in America, 1844–1897*, Cambridge, MA; London: Harvard University Press, 1994, p. 190.
44. Ibid., p. 195.
45. On the growth of national financial market in the US, see, K. G. Garbade and W. L. Silber, 'Technology, Communication and Performance of Financial Markets 1840–1975', *The Journal of Finance*, vol. 33 (3), June 1978, pp. 819–32.
46. Blondheim, *News Over the Wires*, p. 192.
47. Ibid., p. 195.
48. Cf. Graham Storey, *Reuters' Century 1851–1951*, London: Max Parrish, 1951.
49. Roderick Jones, *A Life in Reuters*, London: Hodder and Stoughton, 1951.
50. *The Standard*, 21 July 1894, cited in *Souvenir*, p. 141.
51. *Annual Report of the Telegraph Department 1874–5*, pp. 9–10.
52. *Annual Report of the Telegraph Department 1882–3*, p. 9.
53. Arnold Keppel, *Gun-Running and the Indian North-West Frontier*, London: John Murray, 1911.
54. Hardinge Papers–117, vol. I, *Correspondence 1910–11*, To Viscount Morley, Secretary of State for India, no.32, 18 April 1911.CUL.
55. Ibid., p. 63.
56. George Johnson (ed.), *The All Red Line: The Annals and Aims of the Pacific Cable Project*, London; Ottawa: Edward Stanford, 1903, p. 421.
57. Barty-King, *Girdle Round the Earth*, p. 67.
58. Johnson (ed.), *The All Red Line*, pp. 63–5.
59. Barty-King, *Girdle Round the Earth*, pp. 57–8.
60. Johnson (ed.), *The All Red Line*, p. 126.
61. Ibid., p. 129.
62. Ibid., pp. 133–4.
63. Ibid., pp. 138–9, p. 142.
64. Ibid., p. 140.

65. Ibid., p. 147.
66. Barty-King, *Girdle Round the Earth*, p. 87.
67. John Fisher, *Curzon and British Imperialism in the Middle East 1916–1919*, London: Frank Cass, p.2 95.
68. *Souvenir*, 1894, p. 90. WPL.
69. Ibid., p. 140. WPL.
70. L/PWD/7/No1522, Arms Trade in the Persian Gulf. No. 602–P, from the Director, Persian Gulf Telegraphs, Karachi, to the Director-in-Chief, Indo-European Telegraph Department, Bombay, 5 September 1910. OIOC.
71. G. D. H. Cole and Raymond Postgate, *The Common People 1746–1946*, London: Metheun and Co., 1963 (first edn 1938), pp. 444–6.
72. *Souvenir*, p. 17. WPL.
73. Ibid., p. 89. WPL.
74. Ibid., p. 141.
75. Shannon, *The Crisis of Imperialism*.
76. Cf. Kieve for Britain; Ahvenienen for the Caribbean; Headrick for an overall view: this was by no means a unique Indian experience.
77. *Souvenir*, p. 128. Cf. Jorma Ahvenainen, *The History of the Caribbean Telegraphs before the First World War*, Annales Academiae Scientiarum Fennicae. Ser. Humaniora, The Finnish Academy of Science and Letters, vol. 283, Helsinki, 1996, pp. 41–5.
78. Ahvenainen, *The History of the Caribbean Telegraphs*, p. 32.
79. *Souvenir*, p. 129. WPL.
80. Ibid, p. 140. WPL.
81. *Year Book of Wireless Telegraphy*, 1917, pp. 952–62. GPO.
82. *Europe and America*, p. 39. OIOC.
83. Ibid., pp. 12, 26. OIOC.
84. Ibid., p. 21. OIOC.
85. Ibid., p. 47. OIOC.
86. Ibid., p. 52. OIOC.
87. John Lloyd, *Ode on the Princess Victoria*, 1834 (repr. 1897), London: "Filius", 1897. OIOC.
88. W. H. French, *The Railway Spiritualized: To Which is Added the Electric Telegraph Moralized*, [late Supdt., Submarine Telegraph Company, between England and France, and formerly electrographic tutor to the Electric Telegraph Company], IIIrd revised and enlarged edition, Halesworth; London: James Tippell; Wertheim & Macintosh, 1857, p. 14. OIOC.
89. Ibid., 'The Electric Telegraph Moralized', fn. 'The Earth's Circuit', p. 18. OIOC.
90. Ibid., fn. 'Time Difference', p. 19. OIOC.
91. Cf. Anonymous, *The Story of Cyrus Field: The Projector of the Atlantic Telegraph*, London: T. Nelson & Sons, 1875. OIOC.
92. Munro, *The Wire and the Wave*, p. 18. OIOC.
93. Government of the Union of South Africa, *Speedways of Thought: The Romance of the Development of the Telephone, the Telegraph, of Wireless Telephony and the Airmail*, Pretoria: Government Printer, 1937, Foreword by F. C. Sturrock, Acting Minister of Posts and Telegraphs, p. 9. OIOC.
94. Cf. the frontispiece of this book.
95. Munro, *The Wire and the Wave*, p. 36. OIOC.

96. Smith and Wise, *Energy and Empire*, p. 649.
97. *Souvenir*, p. 14. WPL.
98. Charles Dickens, Letter written on 27 July 1851,in Graham Storey, Kathleen Tillotson (eds), *The Letters of Charles Dickens*, Pilgrim Edition, vol. 6 (1850–2), Oxford: Clarendon Press, 1988, p. 448; also see Robert Henry Scott, 'The History of Kew Observatory, Proceedings of the Royal Society of London', vol. 39, 1885, pp. 37–86, esp. pp. 51–2, for experiments and demonstrations with atmospheric electricity at Kew.
99. *Souvenir*, p. 89. WPL.
100. Cf. Chapter I.
101. http://scienceworld.wolfram.com/biography/Kelvin.html. Last visited 11/01/2006.
102. Cf. M. N. Wise, 'The Flow Analogy to Electricity and Magnetism. I: William Thomson's Reformulation of Action at a Distance', *Arch. Hist. Exact Sci.* 25 (1), 1981, pp. 19–70.
103. http://www.bbc.co.uk/dna/h2g2/A356889.
104. Smith and Wise, *Energy and Empire*, p. 449.
105. Ibid., p. 491.
106. E. A. Guillemin, *Communication Networks*, 2 vols, New York: John Wiley and Sons Inc., vol. I, p. 9, J. N. Mukherjee Collection, GPO.
107. *Souvenir*, p. 89. WPL.
108. Ibid., p. 141. WPL.
109. Richard Shannon, *The Crisis of Imperialism*.

6 Forging a New India in a Telegraph World: Expansion and Consolidation within India

1. General Department, Government of Bombay, vol. 19, no. 759, 1872, Military Department, no. 4074. Memorandum from the Assistant Secretary, Government of India, no. 933, 31 August 1872, to the Officiating Director General, Indian Telegraph Department. MSA.
2. V/24/4289, Administration report of the Indo-European Telegraph Department for 1872–3. No. 432, from J. U. Bateman-Champain, Director, to the Secretary, Government of India, Public Works Department, 19 December 1873. OIOC.
3. Barty-King, *Girdle Round the Earth*, p. 66.
4. General Department, vol. 63, no. 335. Routes for telegraph messages sent out to India. No. 149–T, Government of India, Public Works Department, Telegraphs. Memorandum from Assistant Secretary, Government of India, Public Works Department, 27 March 1871. MSA.
5. V/24/4289, Administration report of the Indo-European Telegraph Department for 1877–8. No. 1279, from J. U. Bateman-Champain, Director, to Secretary, Government of India, Public Works Department, 15 September 1878. OIOC.
6. Government of India, Public Works Department, Civil Works, Telegraph (A) proceedings, May 1875, nos 15–17. Further correspondence regarding the proposed extension of telegraphic communication to the Andamans. No. 16, from Major J. U. Bateman-Champain, Director, Indo-European Telegraph Department, no. 440, 8 December 1874. NAI.

7. Government of India, Civil Works, Telegraph (A) proceedings, October 1880, nos 1–7, no. 1, from Colonel C. J. Merriman, Secretary, Government of Bombay, to the Secretary, Government of India, Public Works Department. NAI.
8. Barty-King, *Girdle Round the Earth*, pp. 138, 141.
9. Government of India, Public Works Department, Civil Works, Telegraph (A) proceedings, July 1900, nos 3–9. Proposed reduction of charges on telegrams between India and Europe. No. 9. From Sir H. Walpole, Under Secretary of State for India, to, the Secretary, Treasury. Nos 1242–99, 1 November 1899. NAI.
10. Government of India, Public Works Department, Civil Works, Telegraph (A) proceedings, November 1884, nos 5–9, no.5, from Colonel J. U. Bateman-Champain, R. E., DG, Indo-European Telegraph Department, to Secretary, Government of India, Public Works Department, no. 546, 30 July 1884. NAI.
11. Foreign Department, Proceedings of the Mekran Telegraph Department, 1863–64, Volume 1A, no. 1127 of 1863. From Lieutenant Colonel Charles Douglas to J. W. S. Wyllie, Under Secretary to Government of India, 25 August 1863. MSA.
12. Foreign Department, Proceedings of the Mekran Telegraph Department, 1863–64, Volume 1A, no. 50 of 1864, from H. I. Walton to the Commissioner in Sindh, Karachi, 23 January 1864. MSA.
13. Foreign Department, Proceedings of the Mekran Telegraph Department, 1863–64, no. 178 of 1864. From Secretary, Indo-European Telegraphs, Lieutenant Colonel W. F. Mariott, to the Commissioner in Sindh, 6 August 1864. MSA.
14. Foreign Department, Mekran Telegraph Proceedings, 1863–64, vol. 1A, no. 9 of 1864. Memorandum from Under Secretary, Government of India, no. 1179, 17 December 1862, 'Intimates that an arrangement under which the services of Mr. Walton for the temporary charge of telegraph, Sindh circle, applied for in August last, are not now necessary'. Copied to the Commissioner in Sindh, 15 January 1863. MSA.
15. Foreign Department, Indo-European Telegraph Department Proceedings, 1869, vol. 2, no. 500 of 1869. From Director, Mekran Telegraph Department, to Secretary, Government of Bombay, 3 June 1869. MSA.
16. Foreign Department, Indo-European Telegraph Department Proceedings, 1869, vol. 2, no. 165 of 1869. From Lieutenant Colonel J. MacDonald, Secretary to Government of Bombay, to Director, Mekran Telegraph Department, 11 June 1869. MSA.
17. Foreign Department, Indo-European Telegraph Department Proceedings, 1869, vol. 2, no. 127 of May 1869. Letter from Colonel Goldsmid, Chief Director, Indo-European Telegraph Department, no. 212, 30 April 1869. MSA.
18. Foreign Department, Electric Telegraph, October 1867, Nos 44–5 (B). From the Telegraph Department to the Financial Department, 20 September 1867, no. 2571. NAI.
19. Cf., for details about Scudamore and the nationalisation scheme, Alan Clinton, *Post Office Workers: A Trade Union and Social History*, London: Allen and Unwin, 1984; C. R. Perry, *The Victorian Post Office: The Growth of a Bureaucracy*, Woodbridge, UK: Boydell Press, 1992.
20. Foreign Department, Indo-European Telegraph Proceedings, 1869, vol. 2, no. 31 of 1869. From John U. Bateman-Champain, Assistant Director in

Chief, Indo-European Telegraph, to the Under Secretary of State for India, enclosing no. 124 of 1869. From Lieutenant Colonel F. J. Goldsmid, Chief Director, Indo-European Telegraphs, to Secretary, Government of Bombay, 22 March 1869. MSA.
21. Proceedings of the Governor General in Council, Public Works Department, no. 13, 9 June 1870. From the Duke of Argyle, India Office, to the Governor General in Council, 31 March 1870, enclosing Reg. no. 25964, General Post Office, 2 April 1869. From Frank Ives Scudamore, Post master general, to the Under Secretary of State for India. MSA.
22. Ibid. MSA.
23. Electric Telegraph, vol. I, 1856–9, Letters from and to the Court of Directors, no. 8, 1859, from Captain Stewart, R. E., to Major General Bowrer, Inspector General of Stores, India, 3 January 1859. MSA.
24. Indo-European Telegraph Department, letters to the Secretary of State for India, 1865, no. 27, from Government of Bombay to the Secretary of State, India, enclosing letter of the Director, Persian Telegraph, 4 July 1865. MSA.
25. Ibid., no. 52, 1865, from Indo-European Telegraph Department to the Secretary of State, India, 11 December 1865. MSA.
26. The term 'Telegraphists' has been used here to mean telegraph experts and I have used it in this sense in the text.
27. Ibid., MSA.
28. Daniel R. Headrick, *The Tentacles of Progress*, pp. 3–16.
29. *Annual Telegraph Department Report 1866–71*, p. 6. NL.
30. *Annual Report of the Telegraph Department 1874–1875*, Simla, 1875, p. 11. From Lieutenant Colonel R. Murray, Officiating Director General, Telegraphs, Simla, 28 September 1875. NL.
31. *Annual Report 1866–1871*, p. 57. NL.
32. Ibid., p. 16. NL.
33. Ibid., p. 15. NL.
34. Ibid., p. 11. NL.
35. Ibid., p. 9. NL.
36. Johnson (ed.), *The All Red Line*, p. 66.
37. Cf., chapter on 1857 and telegraphy, that is, Chapter 2.
38. *The Brahmo Public Opinion*, 22 August 1878. NL.
39. Basil R. E. Labouchardière Papers, 'The Indian Police 1861 to 1947', typed mss, p. 1; J. C. Curry Papers, Box I, 'Memoirs of an Indian Policeman', II vols, vol. I, pp. 3–5. CSAS.
40. General Department, 1873, vol. 76, no. 169. From H. Wellesley, Officiating Under Secretary, Government of India, to the Secretary, Government of Bombay: divulging of telegraphic messages, no. 633 of 1873, 3 February 1873. MSA. Also Home Department, Police proceedings, January 1873, nos 41–2, from H. Wellesley, as above, to all Local Governments on the question of divulging telegraphic messages. NAI.
41. General Department, 1872, vol. 19, no. 169. From Colonel R. Murray, Officiating Director General of Telegraph (India), to Colonel C. H. Dickens, Secretary to Government of India, Public Works Department, no. 101, 13 June 1872. MSA.
42. N. Gerald Barrier, *Banned: Controversial Literature and Political Control in British India, 1907–1947*, Delhi: Manohar, 1976.

43. Anindita Ghosh, Literature, language and print in Bengal, c. 1780–1905, University of Cambridge: Ph.D. Dissertation, 1998.
44. R. T. F. Kirk, *Paper Making in the Bombay Presidency: A Monograph*, Bombay: Official Publications, 1908; D. N. Mookerjee, *Paper and Papier-mache in Bengal: A Monograph*, Calcutta: Bengal Secretariat Book Department, 1908. MSA. Perhaps, these reports reflected the anxiety of the Government to control the material means of circulation and printing during the panic of 1907–8.
45. Bayly, *Empire and Information*.
46. George Pottinger, *Mayo: Disraeli's Viceroy*, Great Britain: Michael Russell, 1990, pp. 186–8; Owen Tudor Burne alleged that the convict had received letters from 'the Patna malcontents, inciting him to commit the deed', fn10. O. T. Burne, *Letters on the Indian Administration of the Earl of Mayo*, printed for 'private research' at the Private Secretary's Office Press, Simla, 1877. Copy in OIOC, Mss EUR D951/17.
47. The *Times of India* quoted in the *Brahmo Public Opinion*, 28 March 1878. NL.
48. Quoted in *The Brahmo Public Opinion*, 25 April 1878. NL.
49. *The Brahmo Public Opinion*, 25 April 1878. NL.
50. *The Brahmo Public Opinion*, 2 May 1878. NL.
51. *The Brahmo Public Opinion*, 18 July 1878. NL.
52. Antonin Artaud, *The Theatre and its Double: Essays by Antonin Artaud*, Montreuil; London: Calder, 1993.
53. Lady Betty Balfour (ed.), *Personal and Literary Letters of Robert, First Earl of Lytton*, 2 vols, vol. II, London: Longman's Green and Company, 1906, pp. 185–6; Lord Lytton to the Queen, Calcutta, 19 December 1879.
54. Ibid., p. 65, Lord Lytton to Sir James Stephen, Simla, 2 July 1877.
55. *The Brahmo Public Opinion*, 25 April 1878. NL.
56. *The Brahmo Public Opinion*, 2 May 1878. NL.
57. *The Brahmo Public Opinion*, 18 July 1878. NL.
58. Balfour, *Personal and Literary Letters of Robert First Earl of Lytton*, p. 18, from Lord Lytton to Mr Disraeli, Simla, 30 April 1876.
59. N3476, *Selection from the Records of the Government of India, Home Department*, no. 12, Calcutta: Government Printing, 1893. MSA.
60. Home Department, Police Proceedings, nos 11–329, February 1893. Papers relating to a Bill to provide for the more effectual surveillance and control of habitual offenders and for certain connected purposes, with notes. NAI.
61. Ibid., no. 928, from the Government of North Western Provinces and Oudh, 7 September 1891. NAI.
62. Fisher, *Curzon and British Imperialism in the Middle East*, p. 295.
63. Hopkirk, *The Great Game*, p. 440.
64. David Gilmour, *Curzon*, London: Papermac, 1995, pp. 89, 167.
65. Denis Judd, *Empire: The British Imperial Experience from 1765 to the Present*, London: Harper Collins, 1996, p. 325.
66. Judd, *Empire*, p. 78.
67. Government of India, Foreign Department, Electric Telegraph 1867–70, Calcutta: Public Works Department Press, 1871. Nos 1–3, November 1868. NAI.
68. Ibid., nos.1–4(B), November 1868. NAI.
69. B. L. Putnam Weale, *Indiscreet Letters from Peking*, London: Hurst and Blackett, 1906, pp. 203–4; cf. Bayly, 'Informing Empire'.

70. K. C. Baglehole, *A Century of Service: A Brief History of Cable and Wireless Ltd. 1868–1968*, London: Cable and Wireless Ltd., 1970, p. 11.
71. Barty-King, *Girdle Round the Earth*, pp. 123–4.
72. Gilmour, *Curzon*, p. 199; Field Marshall Lord Birdwood, *Khaki and Gown: An Autobiography*, London; Melbourne: Ward, Lock and Co., 1941, pp. 93–5.
73. Barty-King, *Girdle Round the Earth*, p. 395.
74. Birdwood, *Khaki and Gown*, p. 103.
75. R. J. Moore, 'Curzon and Indian Reform', *Modern Asian Studies*, vol. 27 (4), October1993, pp. 719–40; Edwin Montagu, 20 December 1918.
76. Curzon papers, vol. 30, no. 3, Letters and Telegrams from Persons in India. From Lieutenant Colonel F. E. Younghusband, Tibetan Mission, Tibet, 1 January 1904, confidential, to Curzon. NAI.
77. Moore, 'Curzon and Indian Reform', p. 724.
78. Arnold Keppel, *Gun-Running and the Indian North-West Frontier*, London: John Murray, 1911, p. 1; Gilmour, *Curzon*, p. 196.
79. Gilmour, *Curzon*, p. 200.
80. Ibid., p. 202.
81. Ibid., pp. 202–3, 296–7.
82. Stephen P. Cohen, 'Issue, Role, and Personality: The Kitchener-Curzon Dispute', *Comparative Studies in Society and History*, vol. 10 (3), April 1968, pp. 337–55; Gilmour, *Curzon*, pp. 290–1.
83. Moore, 'Curzon and Indian Reform', p. 724.
84. Gilmour, *Curzon*, p. 161.
85. Ibid., p. 153.
86. David Dilks, *Curzon in India*, vol. I, London: Rupert Hart-Davis, 1969, pp. 221, 224.
87. *Pioneer*, 7 July 1911; cited in Lovat Fraser, *India Under Curzon and After*, London: Heinemann, 1911, p. 409.
88. Ibid., p. 397.
89. Cohen, 'Issue, Role, and Personality', p. 350.
90. Birdwood, *Khaki and Gown*, pp. 161–4.
91. Cohen, 'Issue, Role, and Personality', p. 354.
92. For a different view of the controversy, cf. Birdwood, *Khaki and Gown*, pp. 162–3, *passim*.
93. Ibid., p. 353.
94. Fraser, *India Under Curzon*, p. 448.
95. Cohen, 'Issue, Role, and Personality', pp. 341–2.
96. Ibid., p. 340.
97. Ibid., pp. 342, 352.
98. Ibid., p. 341.
99. Gilmour, *Curzon*, p. 297.
100. Ibid., p. 298, pp. 310–12, *passim*.
101. *The Recorder*, Calcutta, 12 March 1904, vol. II, no. 11. Editorial. NL.
102. Gilmour, *Curzon*, p. 344.
103. Hardinge papers, 117, vol. I. *Confidential Telegrams to the Secretary of State for India, 1910–11*. From Lord Hardinge to Lord Crewe, no. 53, Simla, 10 August 1911. CUL.
104. Ibid., no. 55, Simla, 17 August 1911. CUL.
105. Fraser, *India Under Curzon*, p. 402.

106. A. J. Field, 'The Magnetic Telegraph, Price and Quantity Data, and the New Management of Capital', *The Journal of Economic History*, vol. 52 (3) June 1992, pp. 401–13.
107. Sir Pelham Grenville Wodehouse, *Carry on, Jeeves*, Harmondsworth: Penguin Books Ltd., 1999 (first published in 1925), pp. 104–5.
108. For a discussion of the main perspectives on time see Alfred Gell, *The Anthropology of Time: Cultural Constructions of Temporal Maps and Images*, Oxford: Berg, 1992.
109. Moitra, *Electric Telegraph ba tarit bartabaha prakaran*, pp. 150–3.
110. *Report of the Telegraph Department for 1860–1*, pp. 19–21. NL.
111. Ibid., Appendix L, General Branch Circular No. 43, from Major C. Douglas, 6 September 1861. Extract from the periodical *Once a Week* for March 1861, p. 273. NL.
112. *Annual Report of the Telegraph Department 1861–2*, p. 19. NL.
113. Cf. Leonard Waldo, 'The Distribution of Time', *Science*, vol. I (23) 4 December 1880, pp. 277–80.
114. Government of India, Public Works Department, Civil Works, Telegraph (A) Proceedings, July 1872, nos 25–9, from Lieutenant Colonel D. G. Robinson, R. E., Director General, Telegraphs (India), to Secretary, Government of India, Public Works Department, 18 June 1872. NAI.
115. Department of Commerce and Industry, Telegraph (A), August 1905, nos 20–5. Adoption of the Standard Time in India and Burma with effect from 1st July 1905.NAI.
116. Department of Agriculture and Revenue (Revenue), Meteorology (A) Proceedings, July 1904, nos 3–4. Note by Sir J. Eliot. NAI.
117. Department of Agriculture and Revenue (Revenue), Meteorology (A) Notes, June 1905, nos 6–35. Note by E. D. Maclagan, 21 January 1905. NAI.
118. Ibid. Note by Denzil Ibbetson, dt. 22 January 1905. NAI.
119. Ibid. Proceedings no. 30. From the Government of Bombay, no. 135P, 26 June 1905. The Port Trust and the Municipality of Karachi adopted resolutions in favour of the adoption of the Standard Time. NAI.
120. Ibid. Proceedings no. 8. From the Government of Bombay, no. 7148, 29 December 1904. NAI.
121. *Bombay Samachar*, 5 February 1907, also, *Mukhbir-i-Islam*, 4 February1907; *Akbar-e-Saudagar*, 5 February 1907: Report on Native Newspapers [RNP], Bombay Presidency, no. 5. For the week ending 2 February 1907. MSA.
122. *Phoenix*, 16 February 1907; RNN, Bombay. No.7. For the week ending 16 February 1907. MSA.
123. See, Bayly, *Information and Empire*, pp. 143, 149, 171–3, 316.
124. *The Leader/The Indian People*, Allahabad, Thursday, 6 May 1909. NAI.
125. *The Times*, 21 July 1894; cited in *Souvenir*, p. 133. WPL.
126. Minto Papers, letters and telegrams, vol. I, no.166, Minto to A. H. L. Fraser, 8 June 1907; M. N. Das, *India under Morley and Minto*, London: G. Allen and Unwin, 1964, p. 18.
127. Ibid., p. 50.

7 The Telegraph General Strike of 1908

1. Work has been done on the railways for example, I. J. Kerr, who looks at the labour employed in constructing the railways; *Building the Railways of the Raj 1850–1900*, Delhi: Oxford University Press, 1997.
2. For example, G. K. Sharma, *Labour Movement in India*, Delhi: People's Publishing House, 1971; Panchan Saha, *Bangla Sramik Andoloner Itihas*, Calcutta: Mukti Publishers, 1972, pp. 12–15; S. Sarkar, *The Swadeshi Movement in Bengal 1903–1908*, Delhi: People's Publishing House, 1973.
3. The argument regarding community consciousness, as opposed to class-consciousness, is made by D. Chakrabarty, see 'Communal Riots and Labour: Bengal's Jute Mill-Hands in the 1890s', *Past and Present*, 91, May 1981, pp. 140–69. For a critique see S. Basu, 'Strikes and "Communal" Riots in Calcutta in the 1890s: Industrial Workers, *Bhadralok* Nationalist Leadership and the Colonial State', *Modern Asian Studies*, 1998, vol. 34, no. 4, pp. 949–83.
4. However, such distinctions cannot be rigidly maintained for the course of the strike. For example, the Bengali Signallers in the Calcutta Telegraph Office did not openly join the strike but the boy peons, often Eurasian, joined the strike of the peons and clerks.
5. Government of India, C & I, *Scheme for the Reorganization of the Indian Telegraph Department*, J. H. LeMaistre, Officiating Under Secretary, Government of India, and I. C. Thomas, Superintendent of the Telegraph and Personal Assistant to the Director General, Indian Telegraph Department, Simla: Government Publication, 1904, p. 12. NAI.
6. Ibid., p. 29. NAI.
7. Ibid., p. 14. NAI.
8. Ibid., pp. 10–11, 22. NAI.
9. *Report of the Telegraph Committee 1906–7*, Appendix, Calcutta: Superintendent, Government Printing, 1907, pp. 86–90. [Henceforth, *Telegraph Report*]. CUL.
10. Ibid. Summary of the recommendations of the Committee. CUL.
11. Department of Commerce and Industry, Telegraph Establishment (A), December 1907, nos 18–20. Sanction to the employment of women as signallers in the Telegraph Department … native women will be eligible for appointment as signallers; if suitable candidates are available. NAI.
12. *Report of the Telegraph Committee 1906–7*, p.43, par. 52 (I). CUL.
13. Ibid., p. 69, par. 91 (I). CUL.
14. Department of Commerce and Industry, Telegraph Establishment (A), February 1909, nos 1–3. Modification of the conditions of employment of women as telegraphists in the Indian Telegraph Department. NAI.
15. For example, *Bandemataram*, Tuesday 16th October 1906. Advertisements and Notices. Some of the other newspapers that began to be published were *Sandhya* and *Karmayogin*. NL.
16. Sunanda Sen, *Colonies and Empire: India 1890–1914*, London: Sangam, 1992, Table 3.4: Council Bills and telegraphic transfers 1890–1913, p. 80.
17. Commerce and Industry, Telegraph Establishment. (A), nos 3–5, May 1908. NAI.

18. Ibid. NAI.
19. Department of Commerce and Industry, Telegraph Establishment. (A), no. 4. From the Director General Telegraphs, to the Private Secretary to the Viceroy, no. 29–T, 29 April 1908. NAI.
20. Sharma, *Labour Movement in India*, pp. 65–6.
21. Department of Commerce and Industry, Telegraph Establishment (A), no. 4. From the Director General, Telegraphs, no. 29–T, 29 April 1908. NAI.
22. *The Panjabee*, 7 December 1907. CSAS.
23. Department of Commerce and Industry, Telegraph Establishment (A), December 1907, nos 1–15. NAI.
24. Ibid. NAI.
25. *Times of India*, Bombay, Thursday, 20 February 1908: Letter to the Editor. MSA.
26. *Amrita Bazar Patrika*, Saturday, 18 January 1908, p. 5, NL; also *Times of India*, Bombay, Thursday, 20 February 1908. MSA.
27. Ibid., Wednesday, 22 January 1908, p. 9. NL.
28. Department of Commerce and Industry, Telegraph Establishment (A), nos 7–10, February 1907. NAI.
29. Department of Commerce and Industry, Telegraph Establishment (A), nos 1–15, December 1907. NAI.
30. Department of Commerce and Industry, Telegraph Establishment (A), nos 27–9, January 1907. Forwarded with Report from the Director General. NAI.
31. Department of Commerce and Industry, Telegraph Establishment (A), nos 7–10, February 1907, no. 8, From Nagpur. NAI.
32. Ibid. No. 9, From Karachi. NAI.
33. Ibid. No.7, From Bombay. NAI.
34. Department of Commerce and Industry, Telegraph Establishment (A), nos 18–20, January 1907. NAI.
35. Ibid., no. 18, from Sir Sydney Hutchinson, Director General of Telegraphs to the Secretary, Government of India, Department of Commerce and Industry, 15 December 1906. NAI.
36. *Times of India*, Friday, 21 February 1908. MSA.
37. *Amrita Bazar Patrika*, Friday, 28 February 1908, p. 5. NL.
38. Ibid., Saturday, 29 February 1908, p. 7. NL.
39. Ibid., Monday, 2 March 1908, p. 4. NL.
40. Ibid., Friday, 6 March 1908, p. 8. NL.
41. Department of Commerce and Industry, Telegraph Establishment (A), nos 3–8. Administrative report of the Indian Telegraph Department for 1907–8 with notes; *Gujarati*, 5 April 1908. NAI.
42. *Amrita Bazar Patrika*, Friday, 13 March 1908, p. 6. NL.
43. *Mahi Kantha Gazette*, 8 March 1908; *RNP*, Bombay, no. 11. For the week ending 14 March 1908. MSA.
44. *Amrita Bazar Patrika*, Thursday, 5 March 1908, p. 6. NL.
45. *Amrita Bazar Patrika*, 25 February 1908, p. 7. NL.
46. *Times of India*, Saturday, 29 February 1908. MSA.
47. *Modern Review*, Allahabad, July 1907. Review of Sister Nivedita (Margaret Noble)'s 'Glimpses of famine and Flood in Eastern Bengal in 1906'. NL.
48. Sen, *Colonies and Empire*, Calcutta, 1992, Table 3.4: Council Bills and telegraphic transfers 1890–1913, p. 80.
49. *Amrita Bazar Patrika*, Monday, 2 March 1908. NL.

50. Department of Commerce and Industry, Telegraph Establishment (A), nos 3–8. Administrative report of the Indian Telegraph Department for 1907–8 with notes, p. 4. NAI.
51. Department of Commerce and Industry, Telegraph Establishment (A), no. 20. Demi-official from G. Rainy, Under Secretary, Government of India, to T. D. Berrington, Director General, Telegraphs, Tel. No. 2522, 9 March 1908, Calcutta. NAI.
52. *Amrita Bazar Patrika*, 6 March 1908, p. 8. NL.
53. *Amrita Bazar Patrika*, 14 March 1908, p. 8. NL.
54. *Bandemataram*, Calcutta, 3 April 1908. NL.
55. *Gujarati*, 5 April 1908; *RNP*, Bombay Presidency, no. 15. For the week ending 11 April 1908. MSA.
56. Ibid. MSA.
57. *Bandemataram*, 6 April 1908. NL.
58. Ibid., 7 April 1908. NL.
59. Ibid., 8 April 1908. NL.
60. Ibid., 9 April 1908. NL.
61. Reprinted. Ibid., 8 April 1908. NL.
62. *Rast Goftar*, 19 April 1908; RNP, Bombay, no. 16, for the week ending 18 April 1908. MSA.
63. Ibid., *Kesari*, 14 April 1908. MSA.
64. *Bandemataram*, 12 April 1908. NL.
65. Ibid., 12 April 1908. NL.
66. Department of Commerce and Industry, Telegraph Establishment (A), no. 4. From the Director General, Telegraphs, forwarding with his remarks the memorial addressed by Henry Barton, late Telegraph Master, no. 29–T, 29 April 1908. NAI.
67. *Amrita Bazar Patrika*, Friday, 21 February 1908, p. 9. NL.
68. For example, ibid., Tuesday, 25 February 1908, p. 7, letter to the editor from a Railway Mail Service Sorter; *Amrita Bazar Patrika*, Thursday, 27 February 1908, p. 10, letter to the editor accusing the Inspector General of degrading behaviour, arbitrary fines and penal transfers of subordinates for trivial offences, from a 'Sufferer'. NL.
69. *Times of India*, Tuesday, 3 March 1908. MSA.
70. Department of Commerce and Industry, Telegraph Estab. (A), no. 20. Demi-official from G. Rainy, Under Secretary, Government of India, to T. D. Berrington, Director General, Telegraphs, Tel. No. 2522, 9 March 1908. NAI.
71. *Bandemataram*, 13 April 1908. NL.
72. *The Statesman*, Calcutta, 12 April 1908. NL.
73. *Bandemataram*, 13 April 1908. NL.
74. *Times of India*, Tuesday, 3 March 1908: memorial by the Bombay Chamber of Commerce. MSA.
75. *Bandemataram*, 13 April 1908. NL.
76. *The Panjabee*, Lahore, 11 December 1908. RNP, Punjab. SL.
77. Department of Commerce and Industry, Telegraph Establishment (A), March 1908, nos 18–25, no. 18. From the Secretary, Bengal Chamber of Commerce, no. 296, 26 February 1908. NAI.
78. Department of Commerce and Industry, Telegraph Establishment (A), March 1908, no. 19. Note by G. Rainy, 3 March 1908. NAI.

79. Department of Commerce and Industry, Telegraph Establishment (A), March 1908, no. 19. Notes by Harvey and Minto, 7 March 1908. NAI.
80. Department of Commerce and Industry, Telegraph Establishment (A), March 1908, no. 20. To the Secretary, Bengal Chamber of Commerce, nos 2544–59, 10 March 1908. NAI.
81. Department of Commerce and Industry, Telegraph Establishment (A), March 1908, no. 23. To the Secretary, Marwari Chamber of Commerce, nos 3208–59, 26 March 1908. NAI.
82. Ibid. From Secretary, Marwari Chamber of Commerce, no. 59, 26 March 1908. NAI.
83. *Oriental Review*, Bombay, 15 April 1908, reprinted from Max's column in the *Capital*. RNP, Bombay. MSA.
84. Department of Commerce and Industry, Telegraph Establishment (A), December 1908, nos 1–3, Orders of the Government of India on the memorials addressed to the Viceroy by the Signalling Staff. NAI.
85. Department of Commerce and Industry, Telegraph Establishment (A), August 1908, no. 5. Extract from the fortnightly report, commissioner, Agra, 7 April 1908. NAI.
86. Confidential Report on the Native Press, Bombay, no. 15. For the week ending 11 April 1908, p. 13. MSA.
87. *Indu Prakash*, Bombay, 7 April 1908, Eng. cols; RNP Bombay no. 15. MSA.
88. *Sind Gazette*, 3 April 1908; RNP Bombay no. 15. MSA.
89. *Sanj Vartaman*, 16 April 1908; RNP, Bombay, no. 16. For the week ending 18 April 1908. MSA.
90. *Bandemataram*, 19 April 1908. NL.
91. *Amrita Bazar Patrika*, 4 June 1909, p. 4. NL.
92. *Bandemataram*, 22 April 1908. NL.
93. For example, *Bandemataram*, 24 April 1908. NL.
94. Elizabeth J. Perry, 'From Paris to the Paris of the East and Back: Workers as Citizens in Modern Shanghai', *Comparative studies in Society and History*, vol. 41, no. 2, April 1999, pp. 348–73.
95. Ibid., p. 370; cites Gerald Dennis Surh, 'Petersburg workers in the 1905; strikes, workplace democracy and the revolution', Ph.D. diss., University of California at Berkeley, 1979.
96. Perry, 'From Paris to the Paris of the East', p. 349.
97. Cf. T. H. Marshall, *Citizenship and Social Class*, Cambridge: Cambridge University Press, 1950.
98. G. D. H. COLE , *The World of Labour: A Discussion of the Present and Future of Trade Unionism*, London: G. Bell and Sons, 1913, pp. 193–204.
99. Ibid., p. 194.
100. Ibid., p. 197.
101. Ibid., p. 199.
102. Ibid., p. 196.
103. Ibid., p. 203. He revised this prediction about Britain in the 1928 edition of the book.
104. *Bandemataram*, 22 August 1906. NL.
105. *Kal*, 19 April 1907; RNP, Bombay, no. 16. MSA.
106. *Bandemataram*, 4 September 1906. NL.

107. *Gujarati*, 5 April 1908; RNP, Bombay, no. 15. For the week ending 11 April 1908. MSA.
108. *Parsi*, 14 April 1907; Ibid. MSA.
109. *Times of India*, Wednesday, 18 March 1908. MSA.
110. See S. K. Sen, *Working Class Movements in India 1885–1975*, Delhi: Oxford University Press, 1993, Chapter II, 'Nationalist movement and the rise of the working class 1885–1919', pp. 23–38, *passim*; Sukomal Sen, *Working Class of India: History of Emergence and Movement 1830–1970*, Calcutta: K.P. Bagchi, 1977; I. M. Reisner and N. M. Goldberg (eds), *Tilak and the Struggle for Indian Freedom*, Delhi: People's Publishing House, 1966.
111. Sharma, *Labour Movement in India*, pp. 65–6.
112. Bryna Goodman, *City and Nation: Regional Networks and Identities in Shanghai 1833–1937*, Berkeley: University of California Press, 1989; R. Chandavarkar, *The Origins of Industrial Capitalism in India: Business Strategies and the Working Classes in Bombay*, 1900–40, Cambridge: Cambridge University Press, 1994.
113. Government of India, Public Works Department, Civil Works, Telegraph (A) proceedings, Aden Defence, June 1899, nos 1 and 73, regarding pay of telegraph subordinates deputed to maintain lines at Aden.
114. Henry Pelling, *A History of British Trade Unionism*, London: Macmillan, 1976 (first edn 1963), pp. 126–7.
115. *Amrita Bazar Patrika*, 14 August 1907. NAI.
116. Ibid., 17 August 1907. NAI.
117. D. K. Lahiri Choudhury, 'India's First Virtual Community and the Telegraph General Strike of 1908', *International Review of Social History*, 48(2003), Supplement, pp. 45–71.
118. S. Sarkar, *The Swadeshi Movement in Bengal 1903–1908*, Delhi, 1973; P. Saha, *Bangla Sramik Andoloner Itihas*, Calcutta 1972; G. K. Sharma, *Labour Movement in India*, Delhi, 1971.

8 *Swadeshi* and Information Panic: Functions and Malfunctions of the Information Order, c. 1900–12

For an expanded theoretical argument of 'panic', 'imagined state', and, the 'entanglement' of both, see D. K. Lahiri Choudhury, 'Sinews of Panic and the Nerves of Empire: The Imagined State's Entanglement with Information Panic, India, c. 1880–1912', *Modern Asian Studies*, 38, 4, 2004, pp. 96–002.

1. *Swadesh* means one's own country. *Swadeshi* meant indigenous production and Boycott implied the non-consumption of foreign goods.
2. Donald Read, *The Power of News: The History of Reuters*, Oxford: Oxford University Press, 1999, p. 63.
3. Ibid., pp. 57, 64, 65–7.
4. Home Department, Judicial Proceedings (B), 19 February 1908, no. 20. Application of the Indian Extradition Act 1903 to Berar. NAI.
5. Hardinge Papers–117, vol. I, *Correspondence 1910–11*, no. 36, from Lord Hardinge, Viceroy, to the acting Secretary of State for India, Viscount Morley, Simla, with enclosures, 11 May 1911. CUL.

6. *Bandemataram*, 14 April 1908. NL.
7. Quoted in the *Panjabee*, Lahore, 16 December 1907. RNP Punjab. SL.
8. Das, *India under Morley and Minto*.
9. For a discussion of the intellectual origins of the movement see C. A. Bayly, *Origins of Nationality in South Asia: Patriotism and Ethical Government in the Making of Modern India*, Delhi: Oxford University Press, 1998.
10. V. Chirol, *Indian Unrest*, London, 1910.
11. V. I. Lenin, *The Heritage We Renounce*, London: NA 1897, p. 74.
12. F. C. Daly, *First Rebels: Strictly Confidential Note on the Growth of the Revolutionary Movement in Bengal*, 1911, Sankar Ghosh (ed.), Calcutta, 1981; J. Campbell Kerr, *Political Trouble in India, 1907–1917*, 1917, Mahadebprasad Saha (ed.), Calcutta, 1973; *Report of the Committee Appointed to Investigate Revolutionary Conspiracies in India* (Sedition Committee) also called the Rowlatt Report, 1918, London, Parliamentary Papers, Reports from Commissioners, Inspectors and others, vol. 8. CUL.
13. Ibid., p. 441. CUL.
14. Das, *India under Morley and Minto*; S. Gopal, *British Policy in India*, Cambridge: Cambridge University Press, 1965; A. Seal, *The Emergence of Indian Nationalism: Competition and Collaboration in the Later Nineteenth Century*, Cambridge: Cambridge University Press, 1968; R. K. Ray, *Urban Roots of Indian Nationalism*, Delhi: Vikas, 1979.
15. S. Sarkar, *The Swadeshi Movement in Bengal 1903–1908*, Delhi: People's Publishing House, 1973.
16. Benoy Jiban Ghosh, *Revolt of 1905 in Bengal*, Calcutta: South Asia Books, 1987; D. Laushey, *Bengal Terrorism and the Marxist Left 1905–1942*, Calcutta: Firma K. L. Mukhopadhyay, 1975.
17. R. K. Ray, *Social Conflict and Political Unrest in Bengal 1875–1927*, Delhi: Oxford University Press, 1984; C. A. Bayly, *The Local Roots of Indian Politics: Allahabad 1880–1920*, Oxford: Clarendon, 1975.
18. S. Das, *Communal Riots in Bengal 1905–1947*, Delhi: Oxford University Press, 1991.
19. R. Chandavarkar, *Imperial Power and Popular Politics: Class, Resistance and the State in India c. 1850–1950*, Cambridge: Cambridge University Press, 1998.
20. R. Chandavarkar, *The Origins of Industrial Capitalism in India: Business Strategies and the Working Classes in Bombay 1900–1940*, Cambridge: Cambridge University Press, 1994.
21. R. J. Popplewell, *Intelligence and Imperial Defence: British Intelligence and the Defence of the Indian Empire, 1904–1924*, London: Frank Cass, 1995.
22. David Arnold, 'The Armed Police and Colonial Rule in South India, 1914–1947', *Modern Asian Studies*, vol. II, no. 1, 1977; David Arnold, *Police Power and Colonial Rule: Madras 1859–1947*, Delhi: Oxford University Press, 1987.
23. D. Dignan, *The Indian Revolutionary Movement in British Diplomacy 1914–1919*, Delhi: Allied Publishers, 1983.
24. J. M. MacKenzie, *Propaganda and Empire: The Manipulation of British Public Opinion 1880–1960*, Manchester: Manchester University Press, 1984.
25. Leonard A. Gordon, 'Portrait of a Bengal Revolutionary', *Journal of Asian Studies*, XVII, February 1968, pp. 197–216.
26. Gopal Halder, 'Revolutionary Terrorism', in Atul Chandra Gupta (ed.), *Studies in the Bengal Renaissance*, Calcutta: National Council of Education, 1958, p. 242;

cited in David M. Laushey, *Bengal Terrorism and the Marxist Left 1905–1942*, Calcutta: Firma K. L. Mukhopadhyay, 1975, p. 7.
27. See *Karmayogin*, Calcutta, Saturday, 24 July 1909, no. 5, 8 Shraban 1316, p. 10, Samadhayi's Case: testimony of the approver S. C. Mullick; also *Karmayogin*, Saturday, 18 September 1909, no. 13, 2 Ashwin 1316, p. 8, search of the Binapani Mess in Dhaka. CSAS.
28. *Karmayogin*, Saturday, 18 December 1909, no. 24, 3 Paush 1316, p. 8, confession in the Haludbari Dacoity Case. CSAS.
29. *Karmayogin*, Saturday, 24 July 1909, no. 5, 8 Shraban 1316, p. 10, Samadhayi's Case: testimony of the approver S. C. Mullick. Apparently he even adopted the name Nandakumar, the hero of Bankim's novel. CSAS.
30. *Bharat Jivan*, Benares, 7 January 1907; RNP, United Provinces, no. 2. For the week ending 12 January 1907. NAI.
31. *Panjabee*, Lahore, 31 July 1907. SL.
32. *Bandemataram*, 16 October 1906. NL.
33. *Citizen*, Allahabad, 29 March 1908; RNP, United Provinces, no. 14. For the week ending 4 April 1908. NAI.
34. *Bharatbani*, Allahabad, 31 March 1908; Ibid. NAI.
35. *Bandemataram*, 7 June 1907, NL; cf. Anonymous, *India's Rebirth: Selections from the Writings of Sri Aurobindo with Notes*, Institut de Recherché Evolutives, Paris, 1993. NL.
36. For an important discussion of Aurobindo's writings, especially in the *Arya*, a journal, which he published since 1914 from Pondicherry, see Sugata Bose, *The Spirit and Form of an Ethical Polity: A Meditation on Aurobindo's Thought*, Sri Aurobindo Memorial Oration, Indian Institute of Management Calcutta, Calcutta: Dimension, 2005.
37. *Hindustan Review*, Allahabad, December 1906. Review of Ernest Piriou's *L'Inde Contemporaine et la Mouvement National*; RNP, United Provinces, no. 5. For the week ending 2 February 1907. SL.
38. RNP, Bombay Presidency. For the week ending 21 September 1907, no. 38. MSA.
39. *Mahratta*, 15 September 1908; RNP, Bombay, no. 38. For the week ending 21 September 1908. MSA.
40. Ibid., *Akhbar-e-Saudagar* and *Jam-e-Jamshed*, 18 September 1908. MSA.
41. *Panjabee*, 31 July 1907. CSAS.
42. *Panjabee*, 17 July 1909. CSAS.
43. *Panjabee*, 20 July 1909. CSAS.
44. *Panjabee*, 22 July 1909. CSAS.
45. John Morley, *Rousseau*, vol. I [1712–78], London: Macmillan and Co., 1910.
46. *The Leader/The Indian People*, Allahabad, 6 May 1909. CSAS.
47. Home Department, Political Deposit, February 1908, no. 1. Dissemination of sedition in the vicinity of military stations. From H. Bradley, Acting Chief Secretary, Government of Madras, to Major General H. B. B. Watkins, Deputy Adjutant General, Southern Army, 4 October 1907, no. 9782. NAI.
48. *The Leader/The Indian People*, 6 May 1909. NL/CSAS.
49. George MacMunn, *The Underworld of India*, London: Jarrolds, 1933, p. 135.
50. M. N. Das, *India under Morley and Minto*, p. 139.
51. *Kavya Sudha Nidhi*, Benares, 10 March 1908; RNP, United Provinces, no. 15. For the week ending 11 April 1908. SL.

52. *Karmayogin*, Saturday, 22 January 1909, no. 29, 9 Magh 1316, p. 7. CSAS.
53. *Rahbar*, Moradabad, 28 May 1908; RNP, United Provinces, no. 22. For the week ending 30 May 1908. SL.
54. Home Department, Police (A), July 1911, nos 243 and 245: includes Jenkins' Note of 18 May 1911; Carlyle's Note of 30 May 1911 and Wilson's Note of 24 May 1911. OIOC.
55. Minto Papers, Minto to Morley, 15 August 1906, roll no. 5, cited in Stephen F. Koss, *John Morley at the India Office 1905–1910*, Yale: Yale University Press, 1969, p. 103.
56. Benoy Jiban Ghosh, *Revolt of 1905 in Bengal*, p. 159.
57. Home Department, Police (A), July 1911, nos 243 and 245: Carlyle's Note of 30 May 1911. OIOC.
58. Home Department, Public (A), November 1907, no. 12. NAI.
59. Home Department, Public (A), January 1906, no. 152. NAI.
60. *Charumihir*, 21 August 1907, cited in Ajit K. Neogy, *Partitions of Bengal*, Calcutta, 1987, p. 261.
61. *Amrita Bazar Patrika*, Thursday, 23 June 1908, p. 8. NL.
62. *Indu Prakash*, 21 November 1908; RNP, Bombay, no. 48. For the week ending 28 November 1908. MSA.
63. Don Dignan, *The Indian Revolutionary Problem in British Diplomacy 1914–1919*, Delhi, 1983, pp. 9–10.
64. Home Department, Public-Political Proceedings, May 1908. nos 112–50. Appendix I. NAI.
65. Popplewell, *Intelligence and Imperial Defence*.
66. Arnold, *Police Power*, p. 230.
67. Home Department, Political Secret Despatches, 15 November 1887, no. 179. From the Marquis of Dufferin and Ava to Viscount Cross. NAI.
68. Home Department, Judicial-Political, March 1901, no. 6. Note by Irwin, General Superintendent of the Thagi and Dakaiti Department, addressed by Government of India to the Secretary of State, 28 March 1901. NAI.
69. Home Department, Police (A), August 1901, nos 83–7. NAI.
70. Home Department, Police (A), February 1904, nos 158–67. NAI.
71. Home Department, Political Deposit, May 1908, no. 1. Formation of a political service under the control of the Director, Criminal Intelligence, and the Criminal Intelligence Department to furnish information about the spread of sedition. Note by C. J. Stevenson-Moore, DCI, 13 June 1908. NAI.
72. Home Department, Political Deposit, May 1908, no. 1. Formation of a political service under the control of the Director, Criminal Intelligence, and the Criminal Intelligence Department to furnish information about the spread of sedition. Notes by H. Adamson, H. H. Risley, Minto, 13 January 1908. NAI.
73. Ibid. Note by H. H. Risley. NAI.
74. Selections from the records of the Government of India, Foreign Department, no. CCCCXLIII, Foreign Dept. serial no. 178, Calcutta, 1910. Correspondence between the Viceroy Lord Minto and certain Ruling Chiefs regarding measures to be taken for the suppression of sedition. From the Nizam of Hyderabad to the Viceroy, 15 October 1909. MSA.
75. Ibid. From the Begum of Bhopal to the Viceroy, 4 September 1909. MSA.
76. Ibid. From the Maharaja of Baroda to the Viceroy, 19 November 1909. MSA.

77. Ibid. From the Raja of Dewas (senior branch) to the Viceroy, 28 September 1909. MSA.
78. Ibid. From the Maharaja of Bikaner to the Viceroy, 29 December 1909. MSA.
79. *Karmayogin*, Saturday, 15 January 1910, no. 28, 2 Magh 1316, p. 9. CSAS.
80. No. 17, from Lord Hardinge, Viceroy, to the Secretary of State for India, Lord Crewe, Calcutta, 9 February 1911, Hardinge Papers–117, vol. I, *Correspondence with the Secretary of State for India 1910–1911*. CUL.
81. MacMunn, *The Underworld of India*, p. 83.
82. No. 5, from J. H. DeBoulay, Secretary to the Viceroy, to S. H. Butler, Member of the Council, Calcutta, 25 November 1910, Hardinge Papers–116, *Letters and Telegrams from Persons in India 1910–11*. CUL.
83. Ibid., p. 7. CUL.
84. *Karmayogin*, Saturday, 14 August 1909, no. 8, 29 Shraban 1316, p. 6; also Karmayogin, Saturday, 18 September 1909, no. 13, 2 Ashwin 1316, p. 7. CSAS.
85. *Karmayogin*, Saturday, 26 June 1901, no. 1, 5 Ashad 1316. CSAS.
86. *Karmayogin*, Saturday, 18 September 1909, no. 13, 2 Ashwin 1316, p. 8, search of the Binapani Mess in Dhaka. CSAS.
87. *Karmayogin*, Saturday, 25 September 1909, no. 14, 9 Ashwin 1316, p. 11. CSAS.
88. Ibid., p. 10; also *Karmayogin*, Saturday, 5 February 1910, 23 Magh 1316, p. 9. Arrest of Jyotindranath Mukherjee and search of the premises in connection with the murder of Shamsul Alam, Criminal Intelligence Department. CSAS.
89. *Amrita Bazar Patrika*, Tuesday, 25 May 1909, p. 7, 'Approvers in trouble': the two approvers denied their statements before the higher court and even denied that any dacoity had taken place. NL.
90. *Karmayogin*, Saturday, 29 June 1909, no. 30, 16 Magh 1316, p. 8. CSAS.
91. *Karmayogin*, Saturday, 5 February 1910, no. 31, 23 Magh 1316, Birendranath Gupta, 19, murders Deputy Superintendent in the Criminal Intelligence Department, Khan Bahadur Moulvi Shamsul Alam. CSAS.
92. *Amrita Bazar Patrika*, Thursday, 3 June 1909, no. 11, p. 8. NL.
93. *Karmayogin*, Saturday, 24 July 1909, no. 5, 8 Shraban 1316, p. 11. CSAS.
94. Hardinge papers, no. 117, *Correspondence with the Secretary of State for India 1910–1911*, Confidential no. 23, to the Earl of Crewe, Secretary of State for India, from Lord Hardinge, 2 March 1911. CUL.
95. The typology is most clearly laid out in Laushey, *Bengal Terrorism and the Marxist Left*, pp. 8–9.
96. For example, *Amrita Bazar Patrika*, Thursday, 3 June 1909, no. 11, p. 9. NL.
97. *Karmayogin*, Saturday, 29 January 1910, no. 30, 16 Magh 1316, Netra Dacoity: the sons of the local Munshi; also Karmayogin, Saturday, 5 February 1910, 23 Magh 1316, p. 9, Jyotindranath Mukherjee was a short-hand writer in the Bengal Secretariat in the department of Finance. CSAS.
98. *Karmayogin*, Saturday, 29 January 1910, no. 30, 16 Magh 1316, Netra Dacoity: the informer was a former household employee. CSAS.
99. Tegart Papers, Box III. CSAS.
100. Intelligence Bureau records, no. 885 of 1912, confidential note from R. B. Hughes Butler, to the Chief Secretary, Bengal, 13 June 1912. I thank Dr S. Das and Mr S. Mukherjee, Indian Police Service for their help. Special Bank archives [Henceforth SB].

101. *Amrita Bazar Patrika*, Tuesday, 25 May 1909, p. 3. NL.
102. *Karmayogin*, Saturday, 9 October 1909, no. 16, 23 Ashwin 1316, p. 12. CSAS.
103. *Karmayogin*, Saturday, 4 December 1909, no. 22, 18 Agrahayana 1316, p. 7. CSAS.
104. *Karmayogin*, Saturday, 5 February 1910, no. 31, 23 Magh 1316, p. 9. CSAS.
105. *Karmayogin*, Saturday, 20 November 1909, no. 20, 4 Agrahayana, 1316, p. 9. CSAS.
106. *Karmayogin*, Saturday, 8 January 1910, no. 27, 24 Paush 1316, p. 7. CSAS.
107. *Karmayogin*, Saturday, 15 January 1910, no. 28, 2 Magh 1316, p. 12. CSAS.
108. *Karmayogin*, Saturday, 18 September 1909, no. 13, 2 Ashwin 1316, p. 7. CSAS.
109. *Karmayogin*, Saturday, 25 September 1909, no. 14, 9 Ashwin 1316, p. 9. CSAS.
110. *Karmayogin*, Saturday, 4 September 1909, no. 11, 19 Bhadra 1316, p. 9. CSAS.
111. *Karmayogin*, Saturday, 26 June 1901, no. 1, 5 Ashad 1316, p. 8. CSAS.
112. *Karmayogin*, Saturday, 3 July 1909, no. 3, (Bengali date missing), p. 9. CSAS.
113. *Amrita Bazar Patrika*, Wednesday, 2 June 1909, no. 118, p. 6. NL.
114. *Amrita Bazar Patrika*, Tuesday, 1 June 1909, no. 117, p.10. NL.
115. *Amrita Bazar Patrika*, Thursday, 3 June 1909, no. 11, p. 7. NL.
116. *Amrita Bazar Patrika*, Tuesday, 1 June 1909, no. 117, p. 10. NL.
117. *The Leader/The Indian People*, Allahabad, Sunday, 18 April 1909, p. 6. NL.
118. *Amrita Bazar Patrika*, Tuesday, 25 May 1909, no. 112, p .7. NL.
119. *Amrita Bazar Patrika*, Wednesday, 2 June 1909, no. 118, p. 6. NL.
120. *Amrita Bazar Patrika*, Saturday, 5 June 1909, no. 120, p. 7. NL.
121. *Karmayogin*, Saturday, 18 December 1909, no. 24, 3 Paush 1316, pp. 7–8. CSAS.
122. *Karmayogin*, Saturday, 20 November 1909, no. 20, 4 Agrahayana 1316, p. 7. CSAS.
123. *Amrita Bazar Patrika*, Thursday, 3 June 1909, no. 11, p. 8. NL.
124. Hardinge papers, no. 117, *Correspondence with the Secretary of State for India 1910–1911*, no. 140. Memorandum from Sir J. P. Hewett, 4 February 1911. CUL.
125. Intelligence Bureau Archives, no. 12 of 1911, from L. F. Morshead, Inspector General, Bengal, to F. C. Daly, Deputy Inspector General, Calcutta, no. 308/XXIV–11, 23 May 1911. SB.
126. Bayly, 'Informing Empire and Nation', p. 197.
127. Popplewell, *Intelligence and Imperial Defence*.
128. Dignan, *The Indian Revolutionary Movement in British Diplomacy*.
129. *Report of the Committee Appointed to Investigate Revolutionary Conspiracies in India (Sedition Committee)*, London, *Parliamentary Papers*, Reports from Commissioners, Inspectors and others, vol. 8, 1918, p. 498. CUL.
130. Government of India, Public Works Department, Civil Works, Telegraph (A) proceedings, May 1892, no. 4, extract from the *London China Express*, 'Telegraphic Extension in China', 4 December 1891. NAI.
131. Government of India, Public Works Department, Civil Works, Telegraph (A) proceedings, February 1893, nos 23–41. Papers regarding the Chinese and Russian telegraph lines and the Rusos-Chinese Convention. No.25, from Gundry, to Under secretary, Foreign affairs. NAI.

132. Government of India, Public Works Department, Civil Works, Telegraph (A) proceedings, January 1894, no. 22, from B. Brenan, Consul at Tientsin, to the IO, no. 27, 20 June 1893. NAI.
133. Government of India, Public Works Department, Civil Works, Telegraph (A) proceedings, June 1897, no. 24, from Sir C. M. MacDonald, Minister at Peking, to the Marquess of Salisbury, Foreign Office, no. 29, 10 March 1897. NAI.
134. K. Asakawa, *The Russo-Japanese Conflict: Its Causes and Issues*, Westminster: Archibald Constable & Co., 1904, pp. 284–5.
135. Ibid., Introduction, p. viii.
136. Cf. Government of India, Public Works Department, Civil Works, Telegraph (A) proceedings, May 1883, nos 1–7. Conference held at Paris to devise measures for the better protection of submarine cables. NAI.
137. Birdwood, *Khaki and Gown*, p. 236.
138. D. Hopkin, 'Domestic Censorship in the First World War', *Journal of Contemporary History*, vol. 5 (4), 1970, pp. 151–69.
139. Government of India, Commerce and Industry, Telegraph (A) proceedings, June 1915, nos 33–78. Establishment of a chain of wireless telegraph stations throughout the British Empire. NAI.
140. Ibid., no. 75. Telegram from Secretary of State for India to Viceroy, 2 November 1914. NAI.
141. Norris, 'A Career in Persia 1894–1930', p. 83. OIOC.
142. Barty-King, *Girdle Round the Earth*, pp. 162, 164.
143. Ibid., p. 163.
144. Ibid., p. 166.
145. Ibid., p. 169.
146. Ibid., pp. 170–1; L/E/3/220. No.41 of 1916, Government of India, Commerce and Industry, Telegraphs, to Austen Chamberlain, Secretary of State for India, 13 October 1916. OIOC.
147. Barty-King, *Girdle Round the Earth*, p. 166.
148. Cf. F. C. Isemonger and J. Slattery (comp.), *Confidential: An Account of the Ghadr Conspiracy 1913–15*, Lahore: Government Printing, 1919. NL.
149. V/24/4292. Administrative reports of the Indo-European Telegraph Department from 1912–13 to 1922–3. Administrative report for 1914–15. OIOC.
150. Ibid., pp. 8–10. OIOC.
151. Ibid., Administrative report for 1915–16, p. 8; Norris, 'A Career in Persia 1894–1930', p. 67. OIOC.
152. Ibid., p. 6. OIOC.
153. L/E/3/220. Department of Commerce and Industry. Letters from India for 1916. No. 13 of 1916, Government of India, C&I, Telegraphs, to Austen Chamberlain, Secretary of State for India, 28 April 1916. OIOC, enclosure no. I: no. 88 T, from C. H. Harrison, Officiating Director General, P&T, to Secretary, Government of India, C&I, 6 August 1915. OIOC.
154. Ibid., letter from the DG, P&T, to Secretary, Government of India, C&I, no. 745-s.t. with enclosures, 27 September 1916. OIOC
155. Ibid., letter from the General Superintendent, Eastern Telegraph Co. Ltd., to the DG, P&T, 18 August 1915. OIOC.
156. G. L. Lawford, L. R. Nicholson et al. (eds), *The Telcon Story, 1850–1950*, London: Telcon, 1950, pp. 90–2.

157. V/24/4292. Administrative report for 1914–15, p. 6. OIOC.
158. Norris, 'A Career in Persia 1894–1930', pp. 84–5. OIOC.
159. V/24/4292. Administrative reports of the Indo-European Telegraph Department, 1915–16 to 1919–20. Administrative report for 1915–16, p. 7. OIOC.
160. Norris, 'A Career in Persia 1894–1930', pp. 85–7; Percy Sykes, *Ten Thousand Miles in Persia*, London: John Murray 1901, (Reprinted 1962). OIOC.
161. L/E/3/219. Government of India, C&I, Telegraphs, from Government of India, C&I, to the Secretary of State for India, no. 4, 1915, 28 January 1915. OIOC.

Conclusion

1. V/24/4284, *Reports of the Indian Telegraph Department 1862–1874/5*, Report for 1872–3, Superintendent of Government Printing, 1873. OIOC.
2. L/PJ/3/686, January 1908, no. 1. From Minto-in-Council, to J. Morley, Secretary of State for India, Calcutta, 16 January 1908. OIOC.
3. Department of Commerce and Industry, Post and Telegraph Establishments (A), September 1918, nos 3–8. Rejection by the Secretary of State of a memorial from Mr H. Barton, General Secretary, Indian Telegraph Association, Limited, praying for redress in connection with internment under the Defence of India rules. NAI.
4. Cf. Aditya Mukherjee, *Imperialism, Nationalism and the Making of the Indian Capitalist Class, 1920–1947*, Sage Series in Modern Indian History, Delhi; London: Sage, 2002.
5. Andrew Porter (ed.), *The Oxford History of the British Empire: The Nineteenth Century*, vol. 3, Oxford, 2001.
6. Pyenson, 'Cultural Imperialism and Exact Sciences Revisited', pp. 103–10.
7. Gyan Prakash, *Another Reason: Science and the Imagination of Modern India*, Princeton: Princeton University Press, 1999, pp. 160–1.
8. I. R. Morus, 'The Nervous System of Britain: Space, Time and the Electric Telegraph in the Victorian age', *The British Journal for the History of Science*, vol. 33 (3), September 2000, pp. 455–75.
9. John Lee, *The Economics of Telegraphs and Telephones*, London: Sir Isaac Pitman and Sons, 1913, pp. 24–5.

Glossary

Akharas	gymnasia
Batta	allowance
Bhakti	religious devotion
Dada	leader or 'big' man
Dak	post
Dal	group
Gaddi	seat or throne
Guru-shishya	teacher and disciple
Hundi	promissory note
Lathiyal	stave wielder
Mistri	artisan labourer
Para	neighbourhood
Patrika	journal
Sabhyata	civilisation
Samiti	association
Ustad	teacher
Ustad-shagird	teacher and disciple
Mohalla	locality or neighbourhood
Mofussil	Suburb or countryside

Bibliography

Archival sources

National Archives of India, Delhi
Department of Agriculture and Revenue, Meteorology (A) Proceedings 1902–10.
Department of Commerce and Industry, Telegraph (A) Proceedings 1904–20.
Department of Commerce and Industry, Telegraph Establishment (A) Proceedings 1904–20.
Foreign Department, Electric Telegraph Proceedings (A).
Foreign Department, Political (A), Proceedings.
Home Department (Public) Proceedings 1830–65.
Home Department, Judicial Proceedings (B).
Home Department, Judicial and Political Proceedings.
Home Department, Police (A) Proceedings 1900–15.
Home Department, Political Deposit (A) Proceedings.
Home Department, Political Secret Despatches.
Home Department, Public-Political Proceedings.
Public Works Department, Civil Works Telegraph (A) Proceedings.
Public Works Department, Telegraph Establishment (A) Proceedings.
Report on Native Newspapers, East Bengal and Assam.
Report on Native Newspapers, Bengal.
Report on Native Newspapers, United Provinces.

Oriental and India Office Collection, British Library, London
Political and Judicial Department Proceedings 1900–12.
Council of India Minutes and Memoranda 1858–1947.
Public Works Department Proceedings.
Department of Commerce and Industry, Telegraph Proceedings 1904–20.
Administrative Reports of the Indo-European Telegraph Department, 1872–1931.
Annual Report of the Indian Telegraph Department 1855–1915.

Maharashtra State Archives, Bombay
Electric Telegraph, 1856–9, Letters to and from the Court of Directors.
Foreign Department, Proceedings of the Mekran Telegraph Department.
Foreign Department, Indo-European Telegraph Department Proceedings.
General Department, Indo-European Telegraph Department Proceedings.
Secretariat Record Office, Indo-European Telegraph Department, Letters to the Secretary of State for India.
Report on Native Newspapers, Bombay Presidency.

Institute of Civil Engineers, London
Minutes of the proceedings of the Institution of Civil Engineers.
Miscellaneous tracts.

Asiatic Society Library, Calcutta
Proceedings of the Asiatic Society 1800–45.

Special Branch Archives, Calcutta
Intelligence Bureau Records 1908–16.

General Post Office Archives, Jadunath Mukherjee Collection, Calcutta
Proceedings of the Society of Telegraph Engineers.
Miscellaneous Tracts.

Official publications

Administrative Reports of the Indian Telegraph Department 1856–1920.
Aitchison, C. U. (comp.), *A Collection of Treaties, Engagements, and Sanads Relating to India and Neighbouring Countries* (3rd edn), Calcutta: Office of the Superintendent of Government Printing, India, 1892–3.
Calcutta Annual Directory, Calcutta, 1834.
Dictionary of National Biography, London, 1895.
Douglas, C, *Report on the Existing State of Telegraph Communication in India and the Projects for its Extension*, Calcutta, 1864.
First Report on the Operations of the Electric Telegraph Department in India from 1 February 1855 to 31 January 1856, Calcutta, 1856.
Government of India, Foreign Department, *Telegraph 1867–70*, Calcutta, 1871.
Kirk, R. T. F., *Paper Making in the Bombay Presidency: A Monograph*, Bombay: Official Publications, 1908.
Mookerjee, D. N., *Paper and Papier-Mache in Bengal: A Monograph*, Calcutta: Bengal Secretariat Book Department, 1908.
Report of the Commission on the Failure of the Bank of Bombay, Parliamentary Papers, 1868–9 (4162) XV, London, 1869.
Report of the Committee Appointed to Investigate Revolutionary Conspiracies in India (Sedition Committee), London, 1918.
Report of the Joint Committee Appointed by the Lords of the Committee of Privy Council for Trade and the Atlantic Telegraph Company to Inquire into the Construction of Submarine Telegraph Cables; Together with Minutes of Evidence and Appendix, Parliamentary Papers, Vol. LXII, London, 1860.
Report of the Telegraph Committee 1906–7, Calcutta: Superintendent, Government Printing, 1907.
Scheme for the Reorganization of the Indian Telegraph Department, Simla, 1904.
Selections from the Records of the Government of India, Home Department, Simla, 1912.
Selections from the Records of the Government of India, Foreign Department, Calcutta, 1910.
Speedways of Thought: The Romance of the Development of the Telephone, the Telegraph, of Wireless Telephony and the Air Mail, Pretoria, 1937.
Supplement to the Gazette of India, Administrative Report of the IETD for 1884–5.

Private papers

National Archives of India
Curzon papers

Oriental and India Office Collection
Arnold papers
Campbell papers
Goldsmid papers
Meston and Young papers
Norris papers

National Library of Scotland, Edinburgh
Minto papers

Glasgow University Library, Glasgow
Kelvin papers

Centre for South Asian Studies Library and Collections, Cambridge
Boileau papers
Curry papers
Labouchardière papers
Tegart papers

Cambridge University Library
Hardinge papers

Newspapers and journals

Amrita Bazar Patrika
Bandemataram
Brahmo Public Opinion
Journal of the Asiatic Society
Journal of the Society of Telegraph Engineers
Karmayogin
Modern Review
Panjabee
The Indian Post and Telegraph Magazine
The Leader/The Indian People
The Recorder
Times of India
Yearbook of Wireless Telegraphy

Printed primary sources and articles before 1930

Adley, C. C., *The Story of the Telegraph in India*, London, 1866.
Anonymous, *Bengal Massacre!* London, n. d (1857?).

Anonymous, *The Late Government Bank of Bombay; its history*, London, 1868.
Anonymous, *The Story of My Life by the Submarine Telegraph*, London, 1859.
Ansted, D. T., *The Bottom of the Atlantic and the First Laying of the Telegraph Cable; Being the Substance of a Lecture Delivered at the Working Men's Association of Guernsey*, Guernsey, 1860.
Arnold, E., *The Marquis of Dalhousie's Administration of British India*, 2 vols, London, 1862.
Asakawa, K., *The Russo-Japanese Conflict: Its Causes and Issues*, Westminster, 1904.
Balfour, George, *Trade and Salt in India Free; reprints (by permission) of Articles and letters from The Times*, London, 1875.
Balfour, Lady Betty (ed.), *Personal and Literary letters of Robert, First Earl of Lytton*, 2 vols, London, 1906.
Barlow, W. H., 'On the Spontaneous Electrical Currents Observed in the Wires of the Electric Telegraph', *Philosophical Transactions of the Royal Society of London*, Vol. 139 (1849), pp. 61–72.
Bennet, E., *The Post Office and its Story*, London, 1912.
Birdwood, Field-Marshal Lord, *Khaki and Gown: An Autobiography*, London, 1941.
Brasher, A., *Telegraph to India: Suggestions to Senders of Messages*, London, 1870.
Bright, Charles, *The life story of Sir Charles Tilston Bright, Civil Engineer: In Which is Incorporated the Story of the Atlantic Cable, and the First Telegraph to India and the Colonies*, London, 1908.
Browne, Edward Granville, *A Year amongst the Persians; Impressions as to the Life, Character, and Thought of the People of Persia, 1887–88*, London, 1926.
Campbell Kerr, J., *Political Trouble in India, 1907–1917*, 1917 (ed.) Mahadebprasad Saha, Calcutta, 1973.
Chirol, V., *Indian Unrest*, London, 1910.
Churchill, W., *The River War: An Account of the Reconquest of the Soudan*, New York, 1933.
Clarke, G. R., *The Story of the Indian Post*, London, 1921.
Cole, G. D. H., *The World of Labour: A Discussion of the Present and Future of Trade Unionism*, London, 1913.
Daly, F. C., *First Rebels: Strictly Confidential Note on the Growth of the Revolutionary Movement in Bengal*, 1911 (ed.) Sankar Ghosh, Calcutta, 1981.
Dutta, Akshoy Kumar, *Directions for a Railway Traveller: Bashpiya ratharohidiger prati upodesh; arthat jahara kaler gari arohon kariya gaman karen, tahader tatsankranta bighna nibaraner upay pradarshan*, Calcutta, 1855.
Europe and America: Reports of proceedings at an Inauguration Banquet given on the 15th April 1864, in commemoration of the renewal by the Atlantic Telegraph Company (after six years) of their efforts to unite Ireland and Newfoundland by means of a submarine cable; and at an Anniversary Banquet given the 10th March 1868, in commemoration of the agreement for the establishment of a telegraph across the Atlantic, on the 10 March 1854, by Mr. Cyrus W. Field, London, n.d.
Framjee, Dosabhoy, *The British Raj contrasted with its predecessors and an inquiry into the disastrous consequences of the rebellion in the North West Provinces upon the hopes of the people of India* [Manager of *The Bombay Times*], London, 1858.
Fraser, Lovat, *India under Curzon and After*, London, 1911.
Goldsmid, John, *Telegraph and Travel: A Narrative of the Formation and Development of Telegraphic Communication between England and India etc*, London, 1874.
Gubbins, M. R., *The Mutinies in Oudh*, London, 1858 (repr. Patna 1978).

Guillemin, Ernst A., *Communication Networks*, 2 vols, New York, 1935.
Hamilton, I. G. J., *An Outline of Postal History and Practice*, Calcutta, 1910.
Isemonger, F. C. and Slattery, J. (comp.), *Confidential: An Account of the Ghadr Conspiracy 1913–15*, Lahore, 1919.
Johnson, George (ed.), *The All Red Line: The Annals and Aims of the Pacific Cable Project*, London and Ottawa, 1903.
Kaye, J. W., *The Life and Correspondence of Charles, Lord Metcalfe* (new and revised edn) Vol. II, London, 1858.
Keppel, Arnold, *Gun-Running and the Indian North-West Frontier*, London, 1911.
Kirkman, Marshall M., *Telegraph and Telephone; The Telegraph, Telephone and Wireless Telegraph Illustrated and Described, Containing a Manual of Practical Lessons for Students and Others*, New York and Chicago, 1904.
Lahiri Choudhury, Dharani Kanta, *Bharat-Bhraman*, Calcutta, 1908.
Lee, John, *The Economics of Telegraphs and Telephones*, London, 1913.
Luke, P. V., 'Early History of the Telegraph in India', *Journal of the Institute of Electrical Engineers*, Vol. XX, 1891, pp. 102–22.
Luke, P. V., 'How the Electric Telegraph Saves India', *Macmillan's Magazine*, Vol. LXXVI, 1897, October 1897, pp. 401–6.
MacMunn, George, *The Underworld of India*, London, 1933.
Maitra, Kalidas, *Electric Telegraph ba taritbartabaha prakaran*, Srirampur, 1855.
Maitra, Kalidas, *Manabdehotatwa or the Human Frame: An Easy and Familiar Introduction to the Principles of Anatomy and Physiology*, Srirampur, 1857.
Maitra, Kalidas, *Refutation of Vidyasagar's Widow Remarriage Plan*, Srirampur, 1855.
Maitra, Kalidas, *The Steam Engine and the East India Railway Containing a History of India, With a Chronological Table of the Indian Princes from Judister down to the Present Time etc.*, Srirampur, 1855.
Maitra, Kalidas, *Three Speeches by H. Pratt, Ramlochan Ghose and K. Maitra*, Calcutta, 1856.
Marshman, J. C., *Memoirs of Major General Sir Henry Havelock*, London, 1860.
Mitra, Rajendralal, Hoernle, A. F. R., Bose, P. N., *Centenary Review of the Asiatic Society 1784–1884*, Calcutta, 1885 (repr. 1986).
Morley, John, *Rousseau*, Vol. I [1712–1778], London, 1910.
Munro, J., *The Wire and the Wave; A Tale of the Submarine Telegraph*, The Boys Own Bookshelf Series, London, n.d.
Murray, E., *The Post Office*, London, 1927.
O'Shaughnessy, W. B., *Instructions Relative to Instruments and Offices for the Indian Telegraph Lines*, London, 1853.
Price Wood, Captain J. N., *Travel and Sport in Turkestan*, London, 1910.
Pritchard, Iltudus Thomas, *The Mutinies in Rajpootana*, London, 1860 (repr. 1976).
Putnam Weale, B. L., *Indiscreet Letters from Peking*, London, 1906.
Rihani, Ameen, *Around the Coasts of Arabia*, London, 1930.
Savarkar, V. D., *The Indian War of Independence*, Delhi, 1909.
Scott, Robert Henry, 'The History of Kew Observatory', *Proceedings of the Royal Society of London*, Vol. 39, 1885, pp. 37–86.
Souvenir of the Banquet and Evening Fete in Celebration of the 25th Anniversary of the Establishment of Submarine Telegraphy with the Far East, London, 1894.

Souvenir of the Inaugural Fete at Mr Pender's 18 Arlington Street to Celebrate the Opening of the Direct Submarine Telegraphic Communication to India on 23rd of June 1870, London, 1870.
Stendhal, Henri-Marie Beyle, *The Telegraph: Being the Second Book of Lucien Leuwen*, 1834–35 (trans.) H. L. R. Edwards, London, 1951.
Stewart, Basil (ed.), Colonel Charles E. Stewart, *Through Persia in Disguise; With Reminiscences of the Indian Mutiny*, London, 1911.
Stewart, Irvin, 'The International Telegraph Conference of Brussels and the Problem of Code Language', *American Journal of International Law*, Vol. 23, Issue 2, April 1929, pp. 292–306.
Sykes, Percy, *Ten Thousand Miles in Persia*, London: John Murray, 1901 (repr. 1962).
The Telegram and Telegrapheme Controversy, As Carried on in a Friendly Correspondence between A. C. and H., both M A's of Trinity College, Cambridge, London, 1858.
Trevelyan, George, *Cawnpore*, London, 1894 (4th edn).
Waldo, Leonard, 'The Distribution of Time', *Science*, Vol. I (23) 4 December 1880, pp. 277–280.
Walker, J. T., 'India's Contribution to Geodesy', *Philosophical Transactions of the Royal Society of London*, Vol. 186, 1895, pp. 745–816.

Printed sources and articles after 1920

Adas, Michael, *Machines as the Measure of Men: Science, Technology and Ideologies of Imperial Dominance*, Ithaca: Cornell University Press, 1989.
Ahvenainen, Jorma, *The Far Eastern Telegraphs: The History of Telegraphic Communications between the Far East, Europe and America before the First World War*, Helsinki: Suomalainen Tiedeakatemia, 1981.
Ahvenainen, Jorma, *The History of the Caribbean Telegraphs before the First World War*, Helsinki: The Finnish Academy of Science and Letters, Vol. 283, 1996.
Anderson, Benedict, *Imagined Communities: An inquiry into the Origins of Nations*, London: Verso (rev. and extended edn), 1991.
Arnold, David, *Police Power and Colonial Rule: Madras 1859–1947*, Delhi: Oxford University Press, 1986.
Artaud, Antonin, *The Theatre and Its Double: Essays by Antonin Artaud* (transl.) Victor Corti, Montreuil; London: Calder, 1993.
Baark, Erik, *Lightning Wires: The Telegraph and China's Technological Modernisation 1860–1890*, London: Greenwood, 1997.
Baber, Z., *The Science of Empire*, Albany: State University of New York Press, 1996.
Baglehole, K. C., *A Century of Service: A Brief History of Cable and Wireless Ltd 1868–1968*, London: Cable and Wireless Ltd, 1970.
Barty-King, Hugh, *Girdle Round the Earth: The Story of Cable and Wireless and its Predecessors to Mark the Group's Jubilee 1929–1979*, London: Heinemann, 1979.
Basalla, George, 'The Spread of Western Science', *Science*, 156, pp. 611–21.
Basalla, George, *The Evolution of Technology*, Cambridge: Cambridge University Press, 1988.
Basu, Subho, 'Strikes and "Communal" Riots in Calcutta in the 1890s', *Modern Asian Studies* 32, 4 (1998), pp. 949–83.

Bayly, C. A. (ed.), *The Raj: India and the British 1600–1947*, London: National Portrait Gallery, 1990.
Bayly, C. A., *Information and Empire: Intelligence Gathering and Social Communication in India 1780–1870*, Cambridge: Cambridge University Press, 1996.
Bayly, C. A., *Origins of Nationality in South Asia: Patriotism and Ethical Government in the Making of Modern India*, Delhi: Oxford University Press, 1998.
Bayly, C. A., *The Local Roots of Indian Politics: Allahabad 1880–1920*, Oxford: Clarendon, 1975.
Bearce, George D., *British Attitudes towards India 1784–1858*, London: Oxford University Press, 1961.
Bhabha, Homi K. (ed.), *Nation and Narration*, London: Routledge, 1990.
Bhattacharya, S., 'Cultural and Social Constraints on Technological Innovation and Economic Development: Some Case Studies', *Indian Economic and Social History Review*, 3 (3), September 1966.
Bismillah, Abdul (ed.), *Bharatenduyugeen Vyanga*, Delhi: Sandarv Prakashan, 1989.
Blondheim, Menahem, *News over the Wires; The Telegraph and the Flow of Public Information in America, 1844–1897*, Cambridge, MA; London: Harvard University Press, 1994.
Bose, Sugata, *The Spirit and Form of an Ethical Polity: A Meditation on Aurobindo's Thought*, Sri Aurobindo Memorial Oration, Indian Institute of Management Calcutta, Calcutta: Dimension, 2005.
Breckenridge, C. A., and van der Veer, P. (eds), *Orientalism and the Postcolonial Predicament*, Delhi: Oxford University Press, 1994.
Buchanan, R. A., 'Institutional Proliferation in the British Engineering Profession, 1847–1914', *The Economic History Review*, New Series, Vol. 38 (I), February 1989, pp. 42–60.
Bud, R., and Cozzens, S. (eds), *Invisible Connections*, Bellingham, SPIE Optical Engineering Press, 1992.
Cain, P. J. and A. G. Hopkins, 'Gentlemanly Capitalism and British Expansion Overseas II: New Imperialism, 1850–1945', *The Economic History Review*, New Series, Vol. 40, I, February 1987, pp. 1–26.
Castells, Manuel, *The Information Age: Economy, Society and Culture*, Vol. I: *The Rise of the Network Society*, Cambridge, MA; Oxford: Blackwell, 1996.
Chakrabarty, D., 'Communal Riots and Labour: Bengal's Jute Mill-Hands in the 1890s', *Past and Present*, 91, May 1981, pp. 140–69.
Chandavarkar, R., *Imperial Power and Popular Politics: Class, Resistance and the State in India c. 1850–1950*, Cambridge: Cambridge University Press, 1998.
Chandavarkar, R., *The Origins of Industrial Capitalism in India: Business Strategies and the Working Classes in Bombay 1900–1940*, Cambridge: Cambridge University Press, 1994.
Chatterjee, Partha (ed.), *Texts of Power: Emerging Disciplines in Colonial Bengal*, London: University of Minnesota Press, 1995.
Chaudhuri, K. N. and Dewey, C. J. (eds), *Economy and Society: Essays in Indian Economic and Social History*, Delhi: Oxford University Press, 1979.
Cipolla, C., *Guns and Sails in the Early Phase of European Expansion 1400–1700*, London: Collins, 1965.
Clinton, Alan, *Post Office Workers: A Trade Union and Social History*, London, 1984.

Coates, Vary T., Finn, Bernard, *A Retrospective Technology Assessment: Submarine Telegraphy; The Transatlantic Cable of 1866*, San Francisco: San Francisco Press Inc., for the George Washington University Program of Policy Studies in Science and Technology, 1979.
Cohen, Stephen P., 'Issue, Role, and Personality: The Kitchener-Curzon Dispute', *Comparative Studies in Society and History*, Vol. 10 (3), April 1968, pp. 337–55.
Cohn, B. S., *Colonialism and Its Forms of Knowledge*, Princeton: Princeton University Press, 1996.
Cole, G. D. H. and Postgate, Raymond, *The Common People 1746–1946*, London: Methuen and Co., 1963 (first edn, 1938).
Das, M. N., *India under Morley and Minto*, London: G. Allen and Unwin, 1964.
Das, S., *Communal Riots in Bengal 1905–1947*, Delhi: Oxford University Press, 1991.
Davis, Clarence B. and Wilburn, Jr, Kenneth E. (eds), with Robinson, Ronald E., *Railway Imperialism*, Contributions in Comparative Colonial Studies No. 26, London: Greenwood, 1991.
Deeloss, J. I., Jarvenpaa, S., and Srinivasan, A. (eds), *Proceedings of the 17th International Conference on Information Systems*, New York: ACM Press, 1996.
Deloche, J., *Transport and Communications in India Prior to Steam Locomotion* (trans. by James Walker) 2 vols, Delhi: Oxford University Press, 1993–4.
Dignan, D., *The Indian Revolutionary Movement in British Diplomacy 1914–1919*, Delhi: Allied Publishers, 1983.
Dilks, David, *Curzon in India*, Vol. I, London: Rupert Hart-Davis, 1969.
Dwyer, John B., *To Wire the World: Perry M. Collins and the North Pacific Telegraph Expedition*, London: Praeger, 2001.
Edney, Matthew H., *Mapping an Empire: The Geographical Construction of British India, 1765–1843*, Chicago; London: University of Chicago Press, 1997.
Elwell-Sutton, L. P., *Persian Oil: A Study in Power Politics*, London, 1955.
Fabian, Johannes, *Time and the Other: How Anthropology Makes Its Object*, New York: Columbia University Press, 1983.
Farah, Caesar E. (ed.), *Decision Making and Change in the Ottoman Empire*, Kirksville, MO: Thomas Jefferson University Press, 1993.
Field, A. J., 'The Magnetic Telegraph, Price and Quantity Data, and the New Management of Capital', *The Journal of Economic History*, Vol. 52 (3) June 1992, pp. 401–13.
Fieldhouse, D. K., '"Imperialism": An Historiographical Revision', *The Economic History Review*, Second Series, Vol. XIV (2), 1961, pp. 187–209.
Fieldhouse, D. K., *Economics and Empire 1830–1914*, London: Widenfeld and Nicholson, 1973.
Fisher, M. H., 'The East India Company's Suppression of the Native Dak', *Indian Economic and Social History Review*, 31 (3) 1994, pp. 319–26.
Fisher, M. H., 'The Office of the Akhbar Nawis: The Transition from Mughal to British Forms', *Modern Asian Studies*, 27 (1) 1993, pp. 45–82.
Fleck, James, 'Selectionism Dominant', Working Paper Series 91/7, Department of Business Studies, Edinburgh University, Edinburgh, 1991.
Frank A., Gunder, *Capitalism and Underdevelopment in Latin America*, New York: Monthly Review Press, 1967.
Gallagher, John, and Robinson, Ronald, 'The Imperialism of Free Trade', *The Economic History Review*, Second series, Vol. VI, no. 1, 1953pp. 1–15.

Gell, Alfred, *The Anthropology of Time: Cultural Constructions of Temporal Maps and Images*, Oxford: Berg, 1992.

Ghosh, Benoy, Jiban, *Revolt of 1905 in Bengal*, Calcutta: South Asia Books, 1987.

Giere, Ronald N., *Explaining Science: A Cognitive Approach*, Chicago: Chicago University Press, 1988.

Gilmour, David, *Curzon*, London: Papermac, 1995.

Goodman, Bryna, *Native Place, City, and Nation: Regional Networks and Identities in Shanghai, 1853–1937*, Berkeley: University of California Press, 1995.

Gopal, S., *British Policy in India*, Cambridge: Cambridge University Press, 1965.

Gordon, Leonard A., 'Portrait of a Bengal Revolutionary', *Journal of Asian Studies*, XVII (February 1968), pp. 197–216.

Gorman, M., 'Sir William O'Shaughnessy, Lord Dalhousie, and the Establishment of the Telegraph System in India', *Technology and Culture*, Vol. 12 (4), 1971, pp. 581–601.

Govind, Vijai, 'The Origin of Electric Telegraphy in India with Special Reference to O'Shaughnessy's Contributions', *Proceedings of the Seminar on Science and Technology in the 18th and 19th Centuries*, Session II, Delhi, 1980, pp. 282–300.

Gupta, Atul Chandra (ed.), *Studies in the Bengal Renaissance*, Calcutta: National Council for Education, 1958.

Habermas, Jurgen, *The Structural Transformation of the Public Sphere* (trans.) Thomas Burger with the assistance of Frederick Lawrence, Cambridge: Polity Press, 1989 (first published in 1962).

Headrick, D. R., 'The Tools of Imperialism: Technology and the Expansion of European Colonial Empires in the Nineteenth Century', *The Journal of Modern History*, Vol. 51, 2, June 1979, p. 231–63.

Headrick, D. R., *The Tentacles of Progress: Technology Transfer in the Age of Imperialism, 1850–1940*, New York: Oxford University Press, 1988.

Headrick, D. R., *The Tools of Empire: Technology and European Imperialism in the Nineteenth Century*, New York; Oxford: Oxford University Press, 1981.

Hobsbawm, E., *The Age of Empire: 1875–1914*, London: Weidenfeld & Nicolson, 1987.

Hopkin, D., 'Domestic Censorship in the First World War', *Journal of Contemporary History*, Vol. 5 (4), 1970, pp. 151–69.

Hopkirk, P., *The Great Game: On Secret Service in High Asia*, Oxford: Oxford University Press, 1990.

Hull, David L., *Science as a Process: An Evolutionary Account of the Social and Conceptual Development of Science*, Chicago: Chicago University Press, 1988.

Hyam, Ronald, *Britain's Imperial Century, 1815–1914* (2nd edn) Basingstoke: Palgrave Macmillan, 1993.

Israel, Milton, *Communications and Power: Propaganda and the Press in the Indian Nationalist Struggle*, Cambridge: Cambridge University Press, 1994.

Jones, Roderick, *A Life in Reuters*, London: Hodder and Stoughton, 1951.

Joshi, P. C. (ed.), *Rebellion: 1857 – A Symposium*, Delhi: Peoples Publishing House, 1957.

Judd, Denis, *Empire: The British Imperial Experience from 1765 to the Present*, London: Harper Collins, 1996.

Kaul, C., 'Reporting the Raj: Reflections on Communications, the British Press, and the Indian Empire', *Contemporary India*, Vol. 3 (3), 2004, pp. 1–18.

Kaul, C., *Reporting the Raj: The British Press and India c. 1880–1922*, Manchester; New York: Manchester University Press, 2003.
Kejariwal, O. P., *The Asiatic Society of Bengal and the Discovery of India's Past 1784–1838*, Delhi: Oxford University Press, 1988.
Kerr, I. J., *Building the Railways of the Raj 1850–1900*, Delhi: Oxford University Press, 1997.
Kesavan, B. S., *History of Printing and Publishing in India*, Vol. 1, Delhi: National Book Trust, 1985.
Kieve, Jeffrey, *The Electric Telegraph: A Social and Economic History*, Newton Abbot: David & Charles, 1973.
Koss, Stephen F., *John Morley at the India Office 1905–1910*, Yale: Yale University Press, 1969.
Kumar, Deepak, *Science and the Raj 1857–1905*, Delhi: Oxford University Press, 1995.
Lahiri Choudhury, D. K., '"Beyond the Reach of Monkeys and Men?" O'Shaughnessy O'Shaughnessy and the Telegraph in India, c. 1836–1856', *The Indian Economic and Social History Review*, Vol. XXXVII (3), July–September 2000, pp. 331–59.
Lahiri Choudhury, D. K., '1857 and the Communication Crisis', in S. Bhattacharya (ed.), *Rethinking 1857*, Delhi: Orient Longman, 2007.
Lahiri Choudhury, D. K., 'Calcutta and Censorship during 1857', in *Burdwan Raj College Brochure*, Bardhaman: Burdwan Raj College, 2008.
Lahiri Choudhury, D. K., 'Communication, Identity, and Choice in 1857', in S. C. Chakrabarty (ed.), *The Uprisings of 1857*, Kolkata: The Asiatic Society, 2009.
Lahiri Choudhury, D. K., 'The Telegraph, Censorship, and "Clemency": Canning during 1857', in *Contemporary Perspectives: History and Sociology of South Asia*, Vol. 3 (1) January–June 2009, pp. 34–54.
Landes, David S., *The Wealth and Poverty of Nations: Why Some Are So Rich and Some So Poor*, New York; London: Little Brown and Co., 1998.
Laushey, D., *Bengal Terrorism and the Marxist Left 1905–1942*, Calcutta: Firma K. L. Mukhopadhyay, 1975.
Lawford, G. L., Nicholson, L. R. (eds), *The Telcon Story, 1850–1950*, London: Telcon, 1950.
Lenin, V. I., *Imperialism, the Highest Stage of Capitalism*, in *Selected Works*, 3 Vols, Vol. 1, Moscow: Progress Publishers, 1977, pp. 169–262.
Lightman, Bernard (ed.), *Victorian Science in Context*, London; Chicago: University of Chicago Press, 1997.
Lindley, Lester G., *The Impact of the Telegraph on Contract Law*, New York; London: Garland Publishing Inc., 1990.
MacKenzie, J. M., *Propaganda and Empire: The Manipulation of British Public Opinion, 1880–1960*, Manchester; New York: Manchester University Press, 1984.
MacLeod, Roy and Kumar, Deepak (eds), *Technology and the Raj: Western Technologies and Technical Transfers to India 1700–1947*, Delhi: Sage, 1995.
Majumdar, Mohinilal, *Early History and Growth of the Postal System in India*, Calcutta: Rddhi, 1995.
Marshall, P. J., 'Western Arms in Maritime Asia in the Early Phases of Expansion', *Modern Asian Studies*, Vol. XIV, 1980, pp. 13–28.
Marshall, T. H., *Citizenship and Social Class*, Cambridge: Cambridge University Press, 1950.

Matthias, Peter (ed.), *Science and Society, 1600–1900*, Cambridge: Cambridge University Press, 1972.

Moore, R. J., 'Curzon and Indian Reform', *Modern Asian Studies*, Vol. 27 (4), October 1993, pp. 719–40.

Morgan, Hiram (ed.), *Information, Media and Power through the Ages: Historical Studies XXII*, Dublin: University College Dublin Press, 2001.

Morus, I. R., 'The Nervous System of Britain: Space, Time and the Electric Telegraph in the Victorian Age', *The British Journal for the History of Science*, Vol. 33 (3), September 2000, pp. 455–75.

Morus, Iwan Rhys, *Frankenstein's Children: Electricity, Exhibition, and Experiment in Early Nineteenth Century London*, Princeton, NJ; Chichester: Princeton University Press, 1998.

Mukherjee, A., *Imperialism, Nationalism and the Making of the Indian Capitalist Class, 1920–1947*, Sage Series in Modern Indian History, Delhi; London: Sage, 2002.

Musson, A. E. (ed.), *Science, Technology and Economic Growth in the Eighteenth Century*, London: Methuen, 1972.

Nair, P. T. (ed.), *Proceedings of the Asiatic Society 1817–1832*, 3 vols, Calcutta: Asiatic Society, 1996.

Pelling, Henry, *A History of British Trade Unionism*, London: Macmillan, 1976 (first edn, 1963).

Perry, C. R., *The Victorian Post Office: The Growth of a Bureaucracy*, Woodbridge: Boydell Press for the Royal Historical Society, 1992.

Perry, Elizabeth J., 'From Paris to the Paris of the East and Back: Workers as Citizens in Modern Shanghai', *Comparative Studies in Society and History*, Vol. 41 (2), April 1999, pp. 348–73.

Popper, K. R., *Objective Knowledge: An Evolutionary Approach*, Oxford: Clarendon Press, 1973.

Popplewell, R. J., *Intelligence and Imperial Defence: British Intelligence and the Defence of the Indian Empire, 1904–1924*, London: Frank Cass, 1995.

Porter, Andrew (ed.), *The Oxford History of the British Empire*, Vol. 3: *The Nineteenth Century*, Oxford: Oxford University Press, 2001.

Pottinger, George, *Mayo: Disraeli's Viceroy*, Great Britain: Michael Russell, 1990.

Prakash, Gyan, *Another Reason: Science and the Imagination of Modern India*, Princeton: Princeton University Press, 1999.

Pyenson, Lewis, 'Cultural Imperialism and Exact Sciences Revisited', *Isis*, Vol. 84 (I), March 1993, pp. 103–10.

Ray, R. K., *Social Conflict and Political Unrest in Bengal 1875–1927*, Delhi: Oxford University Press, 1984.

Ray, R. K., *Urban Roots of Indian Nationalism*, Delhi: Vikas, 1979.

Read, Donald, *The Power of News: The History of Reuters*, Oxford: Oxford University Press, 1999.

Reisner, I. M. and Goldberg, N. M. (eds), *Tilak and the Struggle for Indian Freedom*, Delhi: People's Publishing House, 1966.

Rizvi, S. A. A. and Bhargava, M. L. (eds), *Freedom Struggle in Uttar Pradesh*, Vol. II, Uttar Pradesh: Publications Bureau, 1958.

Robinson, Ronald and Gallagher, John, 'The Imperialism of Free Trade, 1815–1914', *The Economic History Review*, Second Series, Vol. VI (1), 1953, pp. 1–15.

Rodney, W., *How Europe Underdeveloped Africa*, Washington: Howard University Press, 1982.

Rostow, Walter W., *The Stages of Economic Growth: A Non-Communist Manifesto*, Cambridge: Cambridge University Press, 1964.
Rungta, R. S., *Rise of Business Corporations in India 1851–1900*, Cambridge: Cambridge University Press, 1970.
Saha, Panchanan, *Bangla Sramik Andoloner Itihas*, Calcutta: Mukti Publishers, 1972.
Sarkar, S., *The Swadeshi Movement in Bengal 1903–1908*, Delhi: People's Publishing House, 1973.
Sattin, A. (ed.), *An Englishwoman in India: The Memoirs of Harriet Tytler 1828–1858*, Oxford and New York: Oxford University Press, 1986.
Seal, A., *The Emergence of Indian Nationalism: Competition and Collaboration in the Later Nineteenth Century*, Cambridge: Cambridge University Press, 1968.
Sen, S. K., *Working Class Movements in India 1885–1975*, Delhi: Oxford University Press, 1993.
Sen, Sunanda, *Colonies and Empire: India 1890–1914*, Calcutta; London: Sangam, 1992.
Shannon, Richard, *The Crisis of Imperialism 1865–1915*, London: Hart-Davis, Macgibbon, 1974.
Sharma, G. K., *Labour Movement in India*, Delhi: People's Publishing House, 1971.
Smith, Crosbie and Wise, M. Norton, *Energy and Empire: A Biographical Study of Lord Kelvin*, Cambridge: Cambridge University Press, 1989.
Sridharani, Kishanlal, *The Story of the Indian Telegraph*, Delhi: Government Printing, 1953.
Standage, Tom, *The Victorian Internet: The Remarkable Story of the Telegraph and the Nineteenth Century's Online Pioneers*, London: Weidenfeld & Nicolson, 1998.
Stokes, E., 'Late Nineteenth Century Colonial Expansion and the Attack on the Theory of Economic Imperialism: A Case of Mistaken Identity?' *Historical Journal*, Vol. XII, 1969.
Stokes, E., 'Late Nineteenth Century Colonial Expansion and the Attack on the Theory of Economic Imperialism: A Case of Mistaken Identity?' *The Historical Journal*, Vol. XII, 21969, pp. 285–301.
Stokes, E., *The Peasant and the Raj: Studies in Agrarian Society and Peasant Rebellion in Colonial India*, Cambridge: Cambridge University Press, 1978.
Storey, Graham and Tillotson, Kathleen (eds), *The Letters of Charles Dickens*, Pilgrim edition, Vol. 6 (1850–1852), Oxford: Clarendon Press, 1988.
Storey, Graham, *Reuters Century 1851–1951*, London: Max Parrish, 1951.
Taylor, P. J. O. (gen. ed.), *A Companion to the 'Indian Mutiny' of 1857*, Delhi: Oxford University Press, 1996.
Vidal, Gore, *Daily Mirror*, 10 July 2002.
Wesseling, H. L., *Imperialism and Colonialism: Essays on the History of European Expansion*, Contributions in Comparative Colonial Studies No. 32, London: Greenwood, 1997.
Winston, Brian, *Media Technology and Society: A History from the Telegraph to the Internet*, London and New York: Routledge, 1998.
Wodehouse, Pelham Grenville, *Carry on, Jeeves*, Harmondsworth: Penguin Books Ltd., 1999 (first published in 1925).

Unpublished sources

Derbyshire, Ian David, 'Opening Up the Interior: The Impact of Railways on North Indian Economy and Society, 1860–1914', Ph.D. diss., University of Cambridge, 1985.

Ghose, Saroj, 'Introduction and Advancement of the Electric Telegraph in India', diss., Jadavpur University, Faculty of Engineering, Calcutta, 1974.

Ghosh, Anindita, 'Literature, Language and Print in Bengal, c. 1780–1905', University of Cambridge: unpublished Ph.D. diss., 1998.

http://www.bbc.co.uk/dna/h2g2/A356889, last visited 11 January 2006.

http://scienceworld.wolfram.com/biography/Kelvin.html. Last visited 11 January 2006.

Lahiri Choudhury, D. K, 'Communication and Empire: The Telegraph in North India c. 1830–1856', unpublished diss., Jawaharlal Nehru University, Centre for Historical Studies, Delhi, 1997.

Shepard, C. M., 'Philosophy and Science in the Arts Curriculum of the Scottish Universities in the Seventeenth Century', unpublished Ph.D. diss., Edinburgh University, 1975.

Index

Africa (*see also* South Africa) 2, 5, 7, 79, 80, 84, 99, 100, 101, 109, 117–118, 143, 145, 147, 205, 215, 78n
Africans 39, 110, 116
Agency House 14, 68
Agra 11, 22–24, 34–37, 40–41, 53–54, 157, 158, 168, 170–171, 188, 203
Agrarian 48–49, 224
Akshoy Kumar Dutt 7, 224
Allahabad 34, 36, 41, 60, 157, 158, 161, 163, 164, 170, 183, 188, 195, 203
America 180, 185, 187, 191, 192, 204, 214
American 135, 173, 185
Anderson 6, 158
Anglicist 13, 14, 18, 211
Anglo Indian 39, 107, 139–142, 146–147, 149, 167, 171, 188, 189, 201–202, 213
Arabia 1, 84, 86, 90, 215–216
Armed Science 32, 106–107
Asiatic Researches 12–13
Asiatic Society 7, 11–15, 17–18, 27
Assegai 80
Atlantic Cable 87, 89, 109–111
Aurobindo Ghose 162, 185, 194, 199, 201
Australia 82, 97, 99, 106, 112, 115–117, 121, 128, 205, 213
Automation 71, 159, 161

Baghdad 89–90, 92, 107, 129, 132, 204
Bank of Bombay (*see also* Government Bank of Bombay) 57, 131
Barton, Henry 162–164, 169, 172, 176
Bay of Bengal 1, 112
Bayly 5–6, 197
Bennington 70

Birdwood 148
Body politic 123, 125, 141
Bombay 11, 12, 23–24, 33–41, 44, 48, 52, 54, 57–60, 63, 72, 86, 90, 100, 107, 112, 130–132, 136, 148, 152–153, 157–158, 160–161, 163, 165–168, 170, 171–173, 176–178, 180–182, 194, 197, 203, 206–207
Boycott 157–158, 176, 179–185, 201, 214, 249
Bribes 60–61, 91, 93, 95, 102
British Empire 2–7, 79, 82, 85–86, 96, 100, 106–107, 111, 114, 116, 129, 143, 148–149, 179, 205, 215
British Indian Empire 1, 7
Burma 57, 83, 95, 97–99, 152, 158–159, 163, 165, 167, 171, 173, 215
Bushehr 129, 132, 206–8

Cable 1, 4, 7, 19, 20, 49, 53, 73–74, 81, 86–89, 93–94, 97, 105–122, 124–131, 133–134, 136, 137, 143–145, 154, 159, 161, 168, 172, 204–208, 211, 213, 215–216, 218
Cable and Wireless 1, 128, 205, 208
Calcutta 11, 13, 16, 18–20, 22–24, 31, 34–37, 39, 41–43, 46, 54, 58, 60, 63, 69, 83, 94, 95, 116, 132, 136, 141, 152, 157–158, 160–163, 165, 166–172, 175–176, 180, 187, 191, 197, 199, 203
Cama, Bikaji 180, 185, 193
Canada 82, 116, 180, 185, 215
Censorship 15, 37, 39, 44–45, 82, 139, 205, 208
Central Asia 132
Chamber of Commerce 21, 168–171
Chapatis 46, 187
Chatterjee, Partha 12
China 2, 27, 57, 64, 81, 83–84, 97, 131, 136, 143–144, 184–185, 203–204, 218

271

Chinese 83, 111, 144–145, 185, 204
Circulation 2, 6, 8, 13, 43, 46–47, 60, 98, 113–114, 137, 150, 162, 181, 187, 189, 197, 201
Civil War 57, 212
Clerk Maxwell 125, 127, 216, 218
Code/ Codes 41, 43–45, 68, 70, 72–75, 108, 11, 121, 145–146, 149, 159, 195, 205, 213
Commodity/commodities 2, 53, 84, 101, 113
Communication/ communications 1–3, 6–8, 22–23, 27, 31, 33–35, 37, 40–43, 45, 47–50, 52, 55, 57, 60, 70–71, 73, 81, 83–84, 86, 89, 93, 97, 99–102, 105–107, 114, 116, 121, 124, 127, 131, 134, 137–138, 142–143, 145, 148–150, 154, 158, 163, 167, 169–170, 172, 174, 177, 179, 198, 203–204, 207–208, 213–215, 218
Conductor/ Conductors 17, 19, 53, 93, 127
Confidentiality 23, 65, 74
Conspiracy cases 195, 201
Constantinople 82, 89–90, 97, 129–130, 133, 206
Cooper's Hill 65–66
Cotton 57, 59, 81, 131, 158, 174, 212
Criminal Intelligence Department 47, 180, 190–192, 194–197, 200, 202
Crises 2, 5, 8, 32, 114, 128, 204
Crisis 2, 6, 8, 31–32, 37, 40, 42, 48, 52, 85, 110, 119, 122, 128, 153, 164, 168, 179, 208, 213–214, 218
Cromer 143
Curzon 118, 143–149, 162, 181–182, 186, 191, 198, 214

Dal 183
Dalhousie 22, 34–35, 37, 52
Defence of India Act 203, 208, 215
Dervishes 80
Diplomacy 82, 92, 100, 107
Discipline 7, 34, 50–51, 53, 55, 57, 59, 61, 63–65, 67, 69, 71, 73–75, 107, 218

Dominant 2, 12, 21, 79, 81, 147

East India Company 11–13, 16, 22, 34–36, 39–40, 45, 213
Egypt 7, 81, 85–87, 100, 121, 143–145, 177, 185, 192, 197, 205
Electric 1, 17, 20, 22–24, 31, 69, 72, 90
Electrical/ Electrically 4, 17, 19, 25, 63–64, 66, 68, 72–73, 89, 106, 111, 124–125, 127, 134, 136, 168, 216–218
Electrician/ Electricians 73, 85, 98
Electricity 2, 16, 18–19, 26, 28, 64, 68, 74, 80, 97, 108, 124–125, 127, 211, 218
Electronic 2, 3, 5, 105–106, 137, 152, 169, 208, 218
English 13–14, 18–20, 26–27, 31, 47–48, 59, 147, 179, 185–186, 197, 198, 212, 218
English language 142
Entanglement 8
Entropy 125–126
Experiment 35
Extremism 147, 157, 180–183, 190, 203
Extremist 46, 157, 162, 181, 183, 198

Fao 1, 107, 126, 129–130, 206
Far East 57, 85, 129, 131, 205
Field Theory 126–127, 218
Field, C. W. 88, 109, 111, 122
First World War 5, 93, 95, 106, 128, 138, 154, 203, 204, 208,
Formal/ Formally 2, 5, 83, 86, 111, 117–118, 215

Gallagher 5, 79
German Wireless Chain 205
Ghosh, Barindra 189, 200
Gleanings in Science 13
Global 4, 5, 7, 8, 48, 50, 76, 85, 95, 105, 111, 114–115, 120–121, 150, 158, 168, 179, 180, 185, 214–215
Goldsmid 57, 89–96, 98, 100, 133–134
Gossip 45, 115, 121, 141, 211

Great Game   82, 84, 96, 216
Gwadar   1, 91–92, 94, 98, 100, 107, 129

Haig, William   150
Hardinge   116, 149–150, 196–197, 201–203
Hindi   43, 47
Hindustani   26, 41
Historiography   3, 79, 157–158, 175, 179
Horizontal   19, 210, 217
Hundi   42

Identity   74, 158, 178, 192, 196, 214
Imagination   2–3, 7, 68, 75, 118, 125–126, 147, 176, 186–187, 200, 210, 216–218
Imperial   1, 3–7, 15, 49, 79–85, 87–91, 95–96, 99, 101–102, 105–107, 110, 115–119, 121, 125, 128–131, 138, 143–149, 183
Imperialism/ Imperialisms   2–7, 48, 79, 81, 82, 85, 94, 99–102, 110, 117–119, 121–122, 125, 128, 143, 145–146, 179, 208, 215
Imperialist   95, 99, 111, 118, 157, 208, 216
Indian Empire   1–2, 6–7, 129
Indian National Congress   31, 147, 181–183, 213
Indo-European Telegraph Department   65, 89, 95, 99, 108, 118, 129–135, 206–207, 215
Informal   1–2, 5, 7, 81, 91, 95, 98, 101–102, 113, 118, 125, 214, 216, 218
Information / Informational   2–8, 16, 26, 32, 37, 39–40, 42, 44, 45–47, 49, 57–61, 64, 74, 82, 84–85, 95–96, 98, 113–115, 117–118, 120–123, 126, 128, 137–140, 142, 147–150, 153–155, 157, 161, 162, 164, 169, 179, 180–181, 183, 186, 188–192, 194–198, 199–202, 208, 211–215, 217–218
Information Order   3, 8, 128, 179, 190, 214–215

Information Panic   5–6, 8, 32, 46, 57, 120, 122, 128, 153–154, 157, 179, 186, 188, 198
Information rich   47, 211
Institution/institutions   1, 3, 12, 13, 15–16, 18, 63, 66, 71, 113, 119, 127, 148, 153, 161, 190, 212
Institutional / Institutionalisation   1, 15, 17, 50, 52, 72, 80, 108, 109, 119, 195
Institutionalise/ Institutionalised   3, 14–16, 27, 42, 51
Instruments (Telegraph)   19–20, 33, 41, 49, 52, 62, 64, 72, 109, 134–136, 171, 212–213
International/ Internationally/ Internationalist   15, 59, 68, 75, 79, 83–84, 98, 115, 117, 121, 123, 131, 134, 136, 144–145, 150, 153, 161, 175, 177, 179, 184–185, 201–202, 204, 208, 213–216, 218
Internationalism   184
Internet   1–2
Investment / Investments   4, 35, 49, 83–84, 87, 98, 105, 118, 119, 122, 127–128, 157, 211, 215

Jones, William   12, 14
Journal of the Asiatic society   13, 17

Kali / Mahakali / Kalighat   46–47, 49
Karachi   85–86, 88, 92, 107, 129–130, 132–133, 152–153, 157–158, 163, 165, 167–168, 172, 206–207
Karmayogin   194, 201–202
Kelvin (*see also* Thomson, William)   106–108, 124–127
Kitchener   80, 147–9, 243, 265
Knowledge   2, 7, 11–14, 16, 20–21, 25, 28, 38, 40–43, 47–48, 51, 63–64, 75, 80, 90–93, 108–109, 138, 146, 160, 171, 211–213

Laboratory   4, 11, 13–16, 18, 35, 51, 80, 108–110, 125, 135–136, 152, 216

Labour 2, 7, 20, 25, 38, 52, 57, 74, 75, 83, 85–86, 93, 122, 143, 157–158, 160, 166–167, 176–178, 183, 197, 214–215
Labour Mobilisation 158, 178, 214
Language 2, 5, 13, 23, 26–28, 48, 50, 59, 63, 70–71, 73–76, 98, 109, 114, 142, 164, 169, 184, 211–212
Lathiyal 200
Legacy 3, 28
Lenin 4–5
Little Lava Rock 117

Macaulay /Macaulayan/ Macauldean 13–14, 18, 21
Madras 11, 23–24, 34–38, 40, 43, 54, 62–63, 112, 130, 136, 141, 151, 152–153, 158, 160–161, 163, 166, 167–168, 171, 191, 198, 203, 205–207
Madras Time 152–153
Maitra, Kalidas 2, 25, 151
Mekran 90–92, 107, 132–133, 233, 240, 258
Malfunction 2, 179, 212
Mandarin 27
Man-on-the-Spot 5, 80, 146–148, 154–155, 216
Marconi 121, 205, 217
Market / Markets / 4, 48, 57, 59, 105, 115, 120, 123, 139, 154, 168, 194, 202
Mashhad 95–96
Maxim Gun 80
Mechanical 19, 64, 74, 79, 81, 125, 213
Mechanistic 126
Medicine 13, 16
Mesopotamia 90, 92, 215
Messages 7–8, 19, 22–24, 27, 33, 39, 42, 44, 46–47, 49, 54–56, 60–61, 63–64, 68–74, 83, 85–86, 93, 98, 106, 114–115, 120–122, 134, 137–138, 150, 154, 163–164, 166–168, 172, 181, 206–207, 212–213, 218
Messes 177, 184
Metropolis/ Metropolises Metropolitan /

Metropolitanisation 13, 35, 83, 107, 109, 126, 149, 3–4, 6, 15, 21, 50, 75, 79, 80, 107, 109–111, 118, 121, 134, 136, 147, 154, 212, 216, 218
Middle East 132
Military Board 21
Milner 143–145
Mint 16–18, 21
Minto 140, 154, 171, 189, 192, 214
Mirza Abbas Khan 95–96
Mirzapur 23
Misinformation 199
Mobile 1–2, 93, 184
Monopoly / Monopolies 85, 87, 100, 112–114, 121, 130, 133–134, 140
Morley 143, 155, 187, 199, 214
Morse 2, 11, 18, 33, 62–63, 67–68, 72–75, 88, 109, 111, 135
Muscat 86, 91–94, 100
Museum 11, 14–15

Nation state 7, 213, 218
Nation/ Nations 2, 5, 35, 105, 114, 117, 123–124, 167, 201, 213, 218
National 1, 3, 27–28, 31, 35, 48, 113, 124, 130–131, 147, 161, 175–177, 181–184, 208, 213–216
National Legacy 28
Nationalisation 134, 176
Nationalised / Nationalise 27, 114, 133, 158
Nationalising 110
Nationalism / Nationalisms 83, 153, 179, 208, 213–214
Nationalist / Nationalists/ Nationhood 2, 3, 25–26, 46, 121, 154, 157, 166, 175–178, 185–186, 189, 194, 196, 199, 201–202, 208, 214, 218
Nervous System 122–123, 125
Network/ Networks 1–5, 7–8, 11, 34, 37, 42, 45, 47–49, 60, 80, 83–85, 89, 96, 100, 106–107, 113, 115, 118–119, 121, 123, 125, 127, 129, 137, 140, 143, 150, 152, 162, 164, 177, 180, 185, 190, 213
Networks of Circulation 47

Index 275

New Imperialism   5–6, 79, 82, 99, 121–122, 125, 143, 215

Opium   17, 23–24, 38, 42, 61, 64, 131, 172
Orientalist   12–14, 18, 27, 211
Ottoman   83, 90–91, 94–95, 102, 130

Packing   69, 134–135, 213
Panic / Panics / Panicked   2, 5–6, 8, 32, 37, 39, 46, 49, 57, 75, 93, 120–122, 128, 153–154, 157, 170–172, 174, 179, 181, 183, 185–193, 195–199, 201, 203, 205, 207, 208, 214, 218
Patna   34, 140
Pender, John   81–82, 87–88, 106–107, 110–114, 117–119, 122, 125–127, 154
Periphery   79–80, 100, 126, 149
Persia / Persian   1, 13, 26, 41, 43, 57, 84, 86, 88, 89–92, 94–98, 100–102, 107–109, 116, 118, 129–130, 132, 133–135, 140–141, 143, 146, 204, 206–208, 215
Persian Gulf   1, 86, 89, 94, 97, 107–109, 116, 118, 146, 208
Peshawar   11, 22, 34–37, 46, 178, 191
Popplewell   190, 202
Post / Posts   3, 16, 19, 22, 33, 40–41, 43, 54, 56, 61, 70, 95, 110, 116–117, 119, 121, 127, 130, 133–135, 139–141, 146, 152–153, 160, 165–166, 169, 198, 202, 206, 210, 213, 216–217
Post Office   3, 41, 43, 54, 56, 110, 117, 119, 121, 130, 133–134, 139–140, 152, 166, 198, 216
Postal   1, 3, 13–15, 41, 53, 70, 94, 126, 139–140, 153, 166, 168, 171, 180
Press / Presses   8, 15, 39, 44, 46–47, 53–54, 62, 82, 113–115, 120–122, 131, 137, 139–142, 147–151, 153, 161, 163–166, 168–170, 172–173, 177, 179, 181–182, 184–186, 193–196, 198, 201, 203, 205–207, 213–214

Press Messages   8, 115, 120–121
Printing Press / Presses   15, 139, 201
Privacy   23
Propaganda   2–3, 32, 39, 42, 44–45, 47, 81–82, 84–85, 94, 137, 139, 154, 162, 173, 183, 188, 193, 215, 218
Prussia   110
Punjab   23–24, 31, 35, 39–41, 54, 84, 101, 116, 142, 146, 162, 181, 191, 193

Railway/ Railways   2, 15, 22, 25–28, 33, 42, 52, 61, 81, 87–88, 94, 97, 100, 118, 140, 142, 150–151, 153, 166, 168, 175, 197, 202, 204, 216
Rangoon   83, 130, 157–158, 163–164, 167–169, 172, 178, 203
Reason / Reasons/ Reasonably   3, 13, 40, 46, 48–49, 83, 90, 94, 116, 118, 122, 130, 133, 135, 143–145, 147, 152, 166, 169, 171, 181, 186, 190, 201–202, 207–208, 214
Reuters   57, 59, 69, 114–115, 142, 179, 202, 213, 214
Revolutionary   147, 157, 162, 179–184, 186–187, 189–190, 193–197, 201–203, 205, 214–215, 217
Rhodes, Cecil   7, 99–100
Robinson   5, 26, 33, 51, 79–81
Rumour / Rumours/ Rumoured   37, 44–46, 49, 57, 121, 137, 141, 154, 163, 170, 185, 187, 194, 211
Russo Japanese War   185, 204, 215

Said, Edward   14
Salva, Francisco   17
Schwendler, Louis J.   135–6
Science / Sciences   2–4, 6–7, 10–11, 13–14, 16, 18, 21, 25, 27, 32, 47, 50–51, 63–64, 68, 79–81, 85, 88, 106–108, 118, 121–124, 126, 216–217, 218
Scramble for Africa   5
Secrecy   28, 38, 39, 42, 44, 60, 64, 71, 115, 140, 174, 184, 188, 212

Secret / Secrets 22, 43, 44, 68, 141, 161, 163, 169, 182, 184–186, 188, 191–195
Secret Self 184
Sedition / Seditionist 43, 46, 114, 141, 174, 179–180, 188, 192–193, 198–199, 201–203
Seditious 43, 137, 141–142, 180, 182, 190, 193, 197, 199, 201
Service 1, 3, 16, 18, 23, 34–35, 63, 65–67, 84, 98, 113, 118, 123, 129, 137–139, 146, 159, 162–163, 165–166, 169–171, 179, 188, 190–192, 194, 202–203, 208, 213
Sheeb Chunder Nundee 21, 23–24, 61
Siemens 85, 89, 109–110, 129–130, 133, 135, 137
Signal / Signals 26, 32–33, 38, 48, 63–65, 67–68, 73, 74, 111, 126, 138, 146, 154, 165, 170, 187, 203, 213
Signaller/ Signallers / Signalled 20–21, 31, 33, 40, 42–44, 50, 59–65, 67–69, 71–72, 74, 76, 83, 98–99, 125, 136–137, 138, 151, 154, 157–161, 163–165, 167–171, 173–175, 177, 203, 212
Signalling 19, 26, 64, 67, 71, 73, 75–76, 112, 159–160, 163, 165, 168–169, 171, 173, 212
Sindh 54–55, 93, 132, 153
South Africa 7, 99, 101, 116, 143, 145, 147, 205
Space 5–6, 26, 34–35, 37, 40, 45, 50, 64, 73, 75, 80–81, 98, 107–108, 126, 145, 150, 177, 184, 194
Speculation 23, 38, 42, 61, 64, 111, 117
Sri Lanka 1, 130
Standardisation of time 150, 177
Stokes 5, 31
Strike / Strikes 7, 93, 153, 156–160, 162, 165–180, 182, 186, 203, 214
Sub-Imperialism 94, 101–102
Submarine 4, 18, 37, 49, 52, 74, 81, 86–89, 92, 97, 102, 105–112, 116–119, 121–124, 128–129, 136, 211, 216, 218

Suez Canal 112, 143
Surveillance 6–7, 118, 137–140, 142–143, 174, 183, 188, 190, 193, 196
Swadeshi 157–158, 166, 175–176, 179, 180–183, 185, 193, 197, 214
Syed Jamalluddin 46
Symbol / Symbols/ Symbolic/ Symbolically/ Symbolised/ Symbolising 3, 27, 44, 46, 74, 82, 85, 106, 119, 121, 122, 124, 150, 173, 198, 216

Technology 2–4, 6–7, 11, 13, 15, 20–21, 25, 27–28, 32, 42, 46–50, 52–53, 64, 68, 70, 72–75, 79, 80–82, 85, 89, 91, 93–94, 102, 105–106, 108–110, 116, 118–119, 121, 122, 124, 126–128, 134–137, 158–159, 208, 212, 216–218
Technopole/ Technopoles 108, 136
Telegram 2, 27, 31, 37, 39–40, 42, 45, 53–54, 56–57, 69–72, 75, 97, 122, 150, 161, 169, 181, 197, 204, 206, 213
Telegrams 22, 24, 31, 53–56, 60, 63, 70–72, 76, 106, 128, 131, 134, 138, 142, 149–150, 161, 164, 170, 172, 180, 196, 205–207, 212–213
Telegraph Association 159, 163–165, 169, 172–173, 176, 215
Telegraph Princes 95
Telegraph Wire 31, 40, 168
Telegraphers 41, 164, 171
Telegraphy 2, 4–8, 11, 17–18, 21, 25–28, 32–33, 43, 49–52, 67, 70, 75–76, 79, 81–82, 85, 87, 88, 92, 94–95, 98, 101–102, 105–111, 114, 117–119, 121–126, 128–129, 136, 146, 150, 151, 204, 211–212, 214–216, 218
Telephone 106, 128, 131, 169, 177, 213, 216
Thomson, William (*see also* Kelvin) 106, 108, 111
Time 2, 5–6, 11–12, 26, 33–34, 38–39, 44–45, 48, 50, 52, 57–58, 60, 62–64, 67, 70–73, 76, 80–81, 87, 98, 108–109, 119, 120–122,

124, 131, 134, 144–146, 150–153, 158–159, 163, 167, 168, 170, 172–173, 175, 177, 185, 187, 192, 197, 208, 212
Transatlantic Cable   109
Transnational/ Transnationally/ Transnationalism   1–2, 5, 7, 8, 58, 178, 213, 214
Transregional   157, 178
Treason/ Treasonable   43, 46,
Turkey   57, 83–84, 88–89, 107, 134, 140, 192, 207, 215

United States   49, 73
University of Edinburgh   16

Unreasonable / Unreasonably   94, 170
Urdu   41, 43, 47

Vernacular Press   39, 137, 140–142, 203, 213
Vertical   19, 74, 175
Victorian Science   124
Visibility   169, 203

Wireless   1, 74, 106, 121–122, 128, 131, 205, 207–208, 213, 215–217

Yugantar   198, 201